THE APE WITHIN US

by the same author

IN SEARCH OF THE RED APE

THE APE WITHIN US

John MacKinnon

COLLINS
St James's Place, London
1978

William Collins Sons & Co Ltd.
London · Glasgow · Sydney · Auckland
Toronto · Johannesburg

First published 1978

ISBN 0 00 216026 9
Set in Baskerville
Made and Printed in Great Britain by
William Collins Sons & Co Ltd Glasgow

For Guy the Gorilla
who has watched more human behaviour than
any other ape

Contents

Acknowledgements

Apart from expressing my thanks to various ape friends who have permitted me to pry into their domestic affairs, I should like to acknowledge some of my less hairy colleagues to whom I am in debt for contributions to the content of this book or for assistance in the field studies from which the book arises.

Firstly I would like to thank Jane Goodall and Hugo van Lawick for introducing me to the world of wild apes and for a great deal of important early discussion. Others who have been of particular help in undertaking or facilitating field work include Miss Sonia Ivey, Mr Adrien Deschryver, Noel and Alison Badrian, Dr David Chivers, Dr Lim Boo Liat and my Malayan field assistants, Kalang s/o Tot and Ali bin Dramen. I wish to thank the Game and Conservation Departments of Tanzania, Zaire, Malaysia and Indonesia for permission to conduct the various field studies and Professor George Varley, Mr Charles Elton and Professor Geoffrey Harrison for their support in raising funds. I am indebted to the Paton Trust, Poulton Fund, Boise Fund, Leverhulme Trust, Royal Society and Science Research Council for their financial support, also the Institute for Medical Research, Kuala Lumpur, the Lembaga Ilmu Pengetahuan, Indonesia, and the I.R.Z.A.C. in Zaire for their scientific sponsorship.

I should like to give special thanks to my favourite physician, Dr B. E. Juel-Jensen, for his multiple administrations, to my sister Lesley for her attractive drawings and to my wife Kathy for her undeserved support.

Finally and most importantly I would like to thank all those with whom valuable discussion has led to the ideas and opinions that hold this book together. Space and memory prevent me listing all but I would like to mention my gratitude to Professor Niko Tinbergen, Dr Desmond Morris, Dr Vernon Reynolds, Dr Adrian Kortlandt, Professor John Napier, Dr Gil Manley, Dr Peter Andrews, Dr Bernard Woods, Dr Dian Fossey, Dr Sandy Harcourt, Dr Herman Rijksen and Dr Paul Gittins.

John MacKinnon

Illustrations

Drawings

The drawings which appear on chapter headings are by Lesley MacKinnon

Maps

A distribution map for each species appears on the first page of chapters 1-6

In the gibbon map on page 17 the dotted line shows the distribution limits of all gibbon species and the shaded areas show the distribution of the white-handed gibbon.

Introduction

Guy the gorilla sits haughty in his cage, watching the human spectators who pause to gasp and giggle at his formidable bulk. His brow is set in a permanent frown which belies the sad and studious expression of his small dark eyes. For thirty years he has studied the fragile bipeds that have shuffled past at a rate of two million per year. What does he see in their smirking, embarrassed faces that so holds his attention? I join in the crowd and for a few moments Guy and I catch each other's stare and I suddenly realize what it is that links us. Guy recognizes me as an ape as surely as I see a little of myself reflected in his own sensitive face. Through all the clothes, strange talk and fancy ways, Guy can see that apely heritage that man shares with the gorilla – the ape within us.

The idea of man as an ape is nothing new. Even before Darwin's theory of evolution by natural selection taxonomists had recognized that in biological terms man was to be numbered with the living apes. With characteristic modesty man named the order of mammals in which he classified himself as the primates or the highest order. He distinguished the lower primates, which include lemurs, lorises and bushbabies, from the higher primates that include the New World monkeys of South America, the Old World monkeys of Africa and Asia, and man's own superfamily, the apes. It is with the three families of this latter group, the very closest of all man's relatives, that this book is concerned – the siamang and gibbon which make up the lesser apes; the gorilla, chimpanzee, bonobo and orang-utan which are all great apes; and lastly, at the peak of the family tree, man himself and his recent ancestors.

During the last twelve years I have travelled many times to Asia and Africa to learn more about our ape relatives in their natural habitats, one hundred and twenty thousand miles of wanderings, frequently living for months on end in tropical jungles with my subjects. My immediate goal has been to gain a better understanding of these animals, but the real driving force behind my interest has been the desire to gain a better understanding of myself.

The apes are an extraordinarily diverse group of animals. It is impossible to generalize about their size, colour, diet, locomotion or social organization, yet they all bear an unmistakably ape stamp. I have attempted in this book to resolve this elusive apeliness which links our whole superfamily and to see how much of this apeliness we can recognize in ourselves. To do this I must take the reader on a strange journey into the jungles of Malaya to pry into the family affairs of gibbons and siamangs; up the eastern mountains of Zaire to witness the gentle dramas of gorilla group life; through the open woodlands of Tanzania and the swamp forests of the Congo basin to observe the complex, open nature of chimpanzee and bonobo communities; and finally into the magnificent rainforests of Borneo and Sumatra to view the strange, solitary ways of the rare and mysterious orang-utan. All these apes are closely adapted to their own particular environment and to understand their form and habits it is necessary to develop a feel for the forests where they live, the shapes of the trees, the distribution of their food, the other animals sharing their remote domains, and even the subtle jungle sounds and moods.

We shall be concerned with the anatomical and locomotor adaptations of the different apes; how their social organization and courtship patterns suit them to their respective life styles; the extent of their intellectual, communicative and cultural developments; and how they compare in these features with each other and with man. Once we have established man's evolutionary position within the ape complex we can proceed to view our own nature against the whole spectrum of ape behaviour. How much of man is merely ape, how much unique? How and why did such a remarkable species as ours emerge from an ape background? Where are we heading from here?

Current theories about man's nature and origin are based largely on our knowledge of the behaviour of open country baboons which are adapted to the same African bush environment that is thought to have moulded man's own evolution. The northern wolf, too, has come in for serious study because, as a co-operative hunter, he makes his living in the same manner as primitive man. The behaviour of baboon and wolf is certainly relevant to an understanding of man, but the baboon is not an ape; he represents only a *monkey* solution to the African bush environment, man is an *ape* solution. Wolf is merely a *dog* solution to the problems of northern co-operative hunting, man is an *ape* solution. Man is not a monkey, even less is he a dog. Similarities between baboon and man or between wolf and man are only analogies like the similarities between the wings of bats and the wings of butterflies. It is to our closest relatives, the apes, that we must look if we want to build theories about man's beginnings on the firmer ground of homology, or shared origins.

The relevance of ape behaviour to our understanding of man's own origins and man's behaviour has, of course, been recognized for many years, but until recently our knowledge has been very incomplete. The apes are forest animals, difficult to study. It is only now, after many years of extensive field studies, that we are in a position to view the patterns that characterize the whole group to which we belong.

While I claim to have made a major contribution with my own pioneer studies of orang-utans, it is only fair to point out that there are many other students of ape behaviour, and longer, more detailed field studies than my own have been conducted on all the species described in this book. Several of these studies, such as those of George Schaller[1] and Dian

Fossey[2] on gorillas, and Jane Goodall[3] on chimpanzees are well-known to the general public, but other studies by Ray Carpenter[4] and John Ellefson[5] on gibbons, David Chivers[6] on siamangs, David Horr,[7] Peter Rodman,[8] Herman Rijksen[9] and Birute Galdikas[10] on orang-utans, Jorge Sabater Pi,[11] Sandy Harcourt,[12] Alan Goodall and Adrien Deschryver on gorillas, Adriaan Kortlandt,[13] Vernon Reynolds,[14] Noel and Alison Badrian[15] and numerous Japanese primate expeditions[16] on chimpanzees and bonobos, have all made tremendous contributions to our understanding of those animals. I justify my own qualification to write this book with the claim that, having studied all the species myself, I can tackle each from a slightly broader viewpoint than my more specialized colleagues.

When we look at man against his ape background many current theories wobble, some collapse, and we are forced to rewrite much of his early history. But before we can do this we must take a careful look at the living apes themselves, their problems and their solutions. We will start this examination with a trip to the Asian jungles to meet the smallest of our ape relatives, the gibbons.

1-16. Throughout the book supernumerals refer to the works, listed by chapter, in the Bibliography.

Part One

APES

CHAPTER I

Gibbons

MALAYA 1973

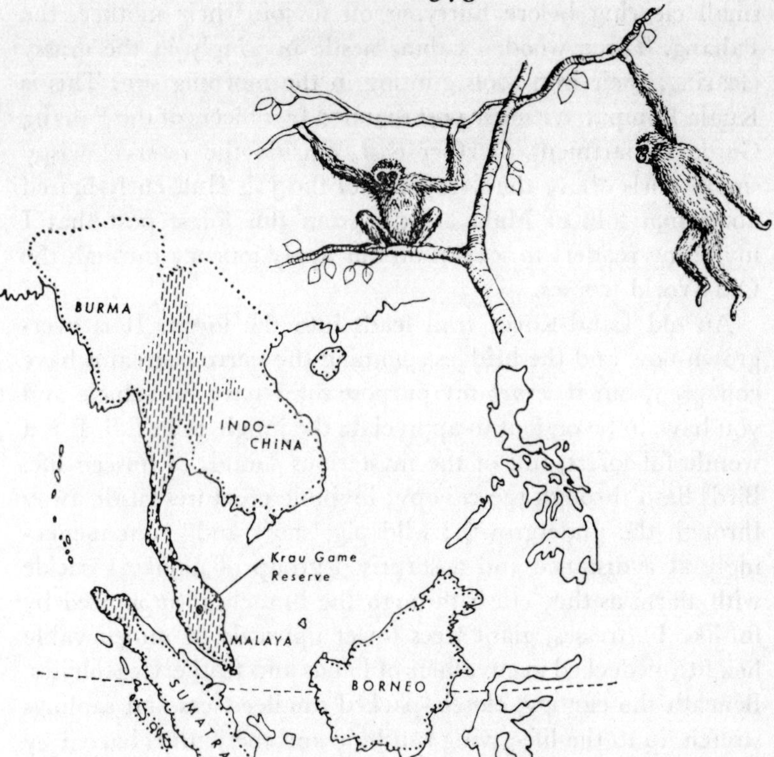

The principal mountain ranges sweep down the Malay Peninsula like a claw springing from the crablike mainland of Asia. Between the open pincers, at the very heart of Malaya, lies the broad valley of the Pahang river, its gentle folds clothed in luxuriant tropical rainforest. The single conical peak of Gunung Benom towers nearly seven thousand feet above the surrounding lowlands, and sheltering under the mountain's eastern slopes is the Krau Game Reserve. Two rivers, the Krau and the Lompat, wind through the tall green forest to meet at a small clearing before hurrying on to join their mother, the Pahang. Three wooden cabins nestle invitingly in the grassy clearing, their iron roofs glinting in the morning sun. This is Kuala Lompat, a ranger post manned by officers of the Pahang Game Department. Further east, outside the reserve, wispy smoke curls above the kampongs of the Jah Hut, curly-haired aboriginal folk of Malaya. It is from this forest post that I invite my readers to set out on our apely journey through the Old World tropics.

An old Land-Rover trail leads into the forest. It is overgrown now and the bridges spanning the narrow streams have collapsed, but it serves my purpose for I have no vehicle and you have to be on foot to appreciate the jungle to its full. It is a wonderful forest, full of the mysterious sounds of unseen life. Birds flash through the canopy; invisible creatures rustle away through the undergrowth; wild pigs snort and grunt menacingly at a distance and a surprised group of monkeys cackle with alarm as they crash through the branches. Supported by fin-like buttresses, giant trees tower upwards to unbelievable heights, bedecked in streamers of lianas and feathery epiphytes. Beneath the canopy, closely-packed smaller trees and saplings stretch up to the life-giving sunshine and the path is barred by prickly rattan palms and aromatic leafy gingers. It is every man's idea of real jungle, with narrow trails twisting through strange silent swamps where the black and white tapirs feed by moonlight. In places the undergrowth is flattened where some great creature has trampled, and huge elephant spoor bear witness to his passing. Sun-bears, leopards and tigers roam deeper in the forest and herds of seladang, the wild forest cattle, make regular pilgrimages to ancient saltlicks. Beneath the shady arches pythons, cobras and weird scaly pangolins go about their secretive business, while up on the slopes of Gunung Benom the rarest inhabitant of the whole jungle survives man's persecution for his magical rhino horns.

It is early morning and the mists are melting in the sparkling shafts of sunlight that pierce the multiple layers of foliage to paint bright patterns on the damp, leaf-carpeted forest floor. Blood-hungry leeches are actively looping towards warm, moving feet, clambering inside wet tennis shoes and feasting on tender ankles. Mosquitoes, too, have picked up the scent of blood and envelop the forest traveller in a persistent whining, dancing haze.

The birds are lustily proclaiming their medley of morning challenge, white-browed bulbul, black-capped jungle babbler, raquet-tailed drongo, crimson-chested bee-eater, fairy bluebirds and gaudy pittas. On the far side of the river a cock argus pheasant gives his beautiful two-toned call, '*Kawau, kawau*', from his dancing ring. High in the treetops a helmeted hornbill replies with his own amazing cackle. Each animal is in his own way defending his rights of territory and space, some singly, some in mated pairs. 'We live here, we are strong, keep clear or else.' The same message in a hundred different languages; one of the oldest messages in the world, the living proof of the competitive struggle for existence, survival of the fittest.

There is another song now, a lively hooting of clear joyful notes. The forest newcomer might be forgiven for imagining it is yet another bird but he would be wrong. The new songster is a mammal and an important one to us, not only a primate but an ape, the white-handed gibbon.

A female gibbon joins in the male's song and their complementary hoots rise and fall in sonorous duet. The male falls momentarily silent while the female performs her unique solo – the gibbon great-call. Three flat, trailing notes rising in pitch, then she launches into an eerie series of prolonged wailing notes which rise to a peak, fall slightly then climb to the crescendo of the call before sailing down again. Now she is rejoined by the hooting male in another excited chorus. Again the female sings her beautiful territorial great-call and this time I can hear the faint echo of an immature daughter joining in behind her mother's wails.

Just ahead is a large strangling fig. The fruit-laden vine has completely smothered the parent tree, taking over most of the crown with its own spreading branches and constricting the host's trunk with a network of descending rootlets, many more than a foot in diameter, as they drop to the ground like the flying buttresses of some distorted cathedral. The helmeted hornbill that was calling earlier is up in the crown of the tree eating figs. He picks a ripe fig clumsily with his heavy beak, rotating and crushing the succulent fruit before tossing it up in the air and catching it in his throat. It is a ludicrously slow and inefficient way of feeding and he drops almost as many fruit as he swallows, but no one could regard this bird as a comic figure. His splendid appearance makes up for his lack of grace. He is a huge black and white bird with two enormously long streamer feathers in his tail and a great red and yellow casque adorning his weighty bill.

The hornbill is not alone in the tree. Two pretty Prevost's squirrels, easily recognized by their bold red, black and white markings, are also making a meal of the fresh figs. I am hoping the gibbons will join them, for the family that is calling nearby is still rather shy of human company. If I follow them when they travel they become nervous and flee faster, but if I remain stationary near a tempting food source they usually tolerate my presence whilst they feed.

The gibbons continue to call but they are definitely travelling this way. I hastily conceal myself in a good position to watch the activities in the fig tree. Just in time! The first gibbon swings through the adjacent trees and climbs straight into the strangling fig without pause. She starts to pluck figs. A second animal, the juvenile female, follows, and finally the adult male moves into the back of the tree by his own personal route. The three tail-less animals circle round the crown greedily plucking only the ripest, reddest fruit, never settling anywhere for more than a few moments, driven by some insatiable restlessness. Gilbert looks like a tiny hairy man, only twelve pounds in weight but superbly coloured, a unique two-toned effect, golden upperparts and dark brown below. Long lateral hair on his cheeks give him a broad triangular head and his impish black face is surrounded by a white fringe. White hairy gloves and socks complete his ensemble. His daughter Gail is a fluffy pale brown gibbon, about half Gilbert's size, while Gertie, the adult female, is chocolate coloured and nearly as large as her mate. On Gertie the white hands and feet and the white face ring show up more clearly and her white extremities flash conspicuously as she brachiates along beneath a branch in the unique acrobatic fashion so typical of gibbons – suspended by her hooked hands and swinging herself forward by alternate pulls of her long, powerful arms.

The gibbons see me, and Gertie gives a few nervous hoots, but they do not stop feeding. Juvenile Gail finds a good spot where there are several red figs within arm's length but before she can finish them Gilbert moves in and Gail is obliged to vacate her site to her father. The animals feed for fifteen minutes then rush out of the tree as suddenly as they arrived. They each leave off the same branch one after the other with a spectacular leap to land in the next tree with hardly a break in their rhythm. I set off after them and keep them in sight for ten minutes but they are too shy to settle and I can no longer keep up with their pace.

Back at the fig tree a party of long-tailed macaque monkeys have discovered the feast and are climbing up casually to feed. The biggest male of the group threatens me, ducking his head and raising his white eyebrow-flashes, while giving harsh, croaking '*kra-kra*' noises. Another gibbon group is calling further to the west. These are Gilbert's neighbours, a family we refer to as group C and characterized by the fact that there are two adult females in the group. The daughter of the family gives full great-calls almost indistinguishable from those of her mother. I set off to look for the new callers.

Over the next two weeks I saw a lot of the local gibbons and especially Gilbert's family. Often I caught only a fleeting glimpse as they dashed off along their leafy pathways, but gradually they were getting more used to my presence and one day I was elated to keep up with them for most of the morning. Usually I was alone, but sometimes my wife, Kathy, came with me. It was hard work; we had to remain constantly alert. A brief wavering of attention and the gibbons were off again and we had lost them. Eventually, when I felt the gibbons were sufficiently tame I allocated Kalang, our field assistant, to the job of following them for a whole day. Kalang was a five-foot aborigine from a nearby village. He had helped with two previous primate studies in the reserve and was already an expert botanist. His father had been a rattan collector in these same forests, so he had learned all the native names of the many trees and vines, and had added to this his new-found knowledge of the habits of the various monkeys and apes. We were trying to make a comparative study of all the forest primates in the area[1] and had already been employing Kalang and his friend Ali for three months to follow the siamangs, the large black cousins of the gibbons. The siamangs were slower than gibbons and had been studied before in detail[2] so they were very tame and easy to follow. We had found that both Kalang and Ali could follow siamangs from dawn to dusk, marking their progress on an accurate map, taking detailed notes every five minutes on the activity of every animal in the family, where and what it was eating. Kalang was the better of the two assistants so I was keen to get him on to the more difficult, shyer gibbons which had not previously been studied in the Krau. I hoped that eventually my wife and I and our two assistants would each be able to follow a different group at the same time to collect synchronous information on the movements of neighbouring families.

I introduced Kalang to the Gilbert family and satisfied myself that he could recognize the different individuals in the group. Next evening Kalang came into camp beaming as usual. Imagine my joy when he laid before me a complete, day-long record of everywhere the family had travelled and what each animal was doing right through the day. Kalang had proved himself indispensable again.

Gibbons turned out to be Kalang's vocation in life and as a tracker of these shy apes he proved supreme. He slipped through the forest like a shadow, never upsetting the gibbons' routine yet faithfully observing and recording everything they did. Once or twice I came upon Gilbert's group but could see no sign of their pursuer. Had Kalang failed to turn up that day, had he gone collecting durian fruit instead, or had his wily prey given him the slip? My fears were groundless. Every evening Kalang would return smiling to our camp and proudly display his notes. Neatly inscribed among the intricacies of gibbon life my own comings and goings were noted down as precisely as those of any other group of monkeys which had crossed paths with Gilbert. Kalang simply melted into the forest and I could easily understand how his tribe had been such successful hunters with the blowpipe in these same shady forests.

For several weeks Kalang and I worked on the gibbons. Ali carried on with the siamang work, while Kathy collected information on the monkey groups. Our picture of Gilbert's family began to build up. Gilbert used a range of about a hundred and twenty acres of forest, which is large for an arboreal primate, and each day he travelled a distance of two kilometres or so around this area. The range of a gibbon family is its territory, an area defended for its exclusive use against all conspecifics. Almost every day the group gave vocal aggressive calls in defence of their living space and spent much of each day patrolling parts of the boundary. Two adult sons, ousted from the tight-knit family group, shared Gilbert's range and both called fairly regularly. George, the elder son, was a dark-reddish animal and he kept well out of the family's way, travelling around the edge of the range. The other young male, Gordon, was a delightful shade of chocolate; he tended to shadow the rest of the family, never too close but never far away. Sometimes if Gilbert and Gertie were feeding in a good food tree Gordon might try to join them but always Gilbert would chase him away.

Almost every evening Gilbert's family returned to the centre of their territory to sleep, choosing their perches from several favoured night trees. The gibbons slept a few yards apart or even in separate trees, each animal hunched tight, sitting on a large horizontal bough.

Like all gibbons they were early risers and six o'clock would see them already travelling towards the edge of their range, feeding as they went. Gilbert himself initiated the family calls and it might be several minutes before Gertie became sufficiently excited to join in with her great-calls. In addition to their territorial message, the calls obviously helped to bind the family together in shared pleasurable activity. Stirred by the duet, neighbouring groups called back and if they sounded too close Gilbert would set off to the extreme edge of his range to re-emphasize his threat and amplify his calls with a visual display. He was a dramatic vision, leaping furiously among the topmost branches of a great tree, hooting for all he was worth. His white face ring, white hands and feet probably enhanced the visual impact of such displays. The great variation in individual coat colour helps make each animal or family of animals very distinctive, so that their presence is not easily ignored by trespassers.

Gilbert's territory was bordered to the south by a grassy clearing and the Lompat river. Other gibbon groups called each day from across the river but Gilbert faced no real risk of territorial encroachment from that direction and left that boundary relatively unpatrolled. To the north a hill ridge served as a barrier, and although there was a family of pale gibbons living on the other side they rarely gave Gilbert cause for concern. To the south-west, Gilbert's eldest son George provided a sort of buffer between Gilbert and any other family. The forest to the east held no family threat but it was occupied by a single male, Stanley, who called quite regularly. Gilbert's most serious challenge at the moment came from the west where there was no real barrier between himself and group C, which consisted of a mated pair with an adult daughter and a juvenile. It was along this boundary that Gilbert concentrated his territorial energies each morning, and once or twice he was involved in actual fights and chases backwards and forwards with his male counterpart in group C.

Territorial calling was very much a morning activity; most of the gibbons had completed their singing by nine o'clock and by midday the forest was quiet. Most gibbon families called at least once a day, often twice and sometimes three times, each bout lasting about twenty minutes. Between choruses the gibbons divided their time equally between travelling and feeding. Much of their travel was to and from food trees or looking for fruit. Fruits, and figs for choice, were important to the gibbons but the apes also consumed large quantities of leaves and even insects and caterpillars. Water was no problem; most of their liquid needs were met by the succulent vegetation or a quick scoop of rainwater caught in a convenient tree-hole. They managed to combine their foraging efficiently with patrolling parts of their range, sometimes retracing their routes of the previous day, although they already knew there was little food along that path.

The gibbon day is a hurried and busy affair. They seem to suffer from the proverbial ants in their pants and will hardly settle for a moment. It is not merely in comparison with the poor biped trying to follow them on the ground that they seem so active. Other forest primates, including the large siamangs, all pursue life at a more leisurely pace, with long periods of rest interspersed among their foraging activities. Not the Gilbert family. Just when you think they have settled in a fruit tree for a good meal they are rushing away again after barely sampling the goods and when, in the heat of midday, they do eventually grab a few minutes' rest, you must watch them like hawks if you are not to lose them as they dash off. The afternoon is almost as hectic as the morning. In the course of a day they may visit sixteen or more well-spaced food trees as well as snatching odd leaves and shoots to nibble as they pass. But suddenly, at about four o'clock, the gibbons slow down. It is as though the clockwork has run down. They make their way to the nearest suitable night tree and by half-past four their day is done. There are still another two hours of daylight but the gibbons are finished, lounging about in their night positions, grooming and sunning in the orange rays of the evening sun, and recharging their batteries for the next day's hectic activity. Why are they in such a hurry all day? Why the rush: the worry that the day won't be long enough? The answer remains a mystery.

The suspensory arm-dominated locomotion of gibbon travel is in direct contrast to the quadrupedal locomotion of arboreal monkeys. Monkeys run along the top of branches using their long tails for balance. They leap from one branch to another with powerful springs of the hind legs. Gibbon brachiation has been interpreted by other authors as primarily a feeding adaptation, to enable them to hang effortlessly whilst plucking food from terminal twig ends among flimsy branch supports. Brachiation has been regarded as inefficient for long-distance travel. I cannot agree with this view. Gibbon travel may indeed cost more energetically per distance covered when compared with quadrupedal monkey locomotion, but it is certainly effective. Not only are gibbons the fastest tree-top travellers, they are also almost silent in comparison to the noisy crashings of monkeys as they leap from one tree to the next. Gibbons occupy the largest home ranges and consistently travel more than twice as far each day as any of the other tree-living primates in the Malayan forest.

As we followed the gibbons more often, their individual personalities began to emerge. Gilbert was clearly an old male, about twenty-five years of age but still full of life. With his strange two-toned coat he was one of the most conspicuous and amusing primate personalities in the forest. He did not care much for his two sons. He saw them as a threat to his own status as the big male of the territory and he rarely tolerated them anywhere near him. He didn't seem all that fond of Gertie or daughter Gail either. Gilbert was for ever rushing off on little forays of his own, often a hundred yards or more away from the other animals. He seemed to enjoy his moments of peaceful solitude. Not that he was in any way a subdued character. Far from it. Gilbert was quite a cheeky gibbon, pestering the larger siamangs with whom he shared part of his range. Again and again the big male siamang, Murgatroyd, chased Gilbert out of a good fig tree until, exasperated by his repeated exercise, he was forced to give in and allow the smaller upstart to feed unmolested in a corner branch. Gertie and Gail watched these proceedings jealously, not daring themselves to enter the tree.

Gertie was the stable element in the family. She steered a less erratic course on their daily travel and Gail stayed close to her. Gertie's long, black, pendulous nipples told a tale of many young gibbons successfully reared, and she seemed still to be a central figure holding the family group together. It was she who gave the thrilling great-calls during the territorial choruses. Young Gail was a small fluffy gibbon. She had just begun to join in the family calls, her squeaky little voice punctuating her parents' duet. She was still quite young and often played on her own, dangling about, biting twigs or jumping and swinging round and round the same branch, spinning like a top.

From what we saw of other gibbons, Gilbert's family seemed absolutely typical. All the families consisted of a single mated adult pair accompanied only by immature young. The six families we knew well averaged only three to four animals and each group occupied an exclusive territory which they were at pains to defend, usually only vocally but with force if necessary. Young were born at approximately three-year intervals and stayed with their parents for eight or more years before being forced or lured away. There were several lone males in the population, either living inside their parents' ranges, as were Gilbert's two sons, or occupying spaces between family territories, like the lone male Stanley. These solitary males called on their own but we never heard lone females singing. It seemed likely that lone males could attract single females or lure mature daughters from their parental territories to set up new breeding units, but we could only speculate as to how neighbouring families would react to the establishment of the new group.

It was 4 April and we had been working in the Krau Reserve for three months. I lay on the hard wooden bed gazing up at the medley of insects that were trapped in the top of our large mosquito net. Kathy was asleep beside me and our six-month-old son, Jamie, was just stirring in his bamboo cot in the corner of the room. The soothing sounds of the Malayan dawn filtered into the wooden cabin: the lusty song of the bulbuls, the gurgling of the river that ran past the game post, and the scufflings of the banded leaf monkeys as they left their lofty sleeping positions. From deeper in the forest came the first clear hoots of Stanley, the lone male gibbon who lived in the eastern corner of the Reserve. I climbed out of bed and dressed quickly. Was there time for something to eat? Or should I make sure of locating him now while he was still calling? I found my compass and took an accurate bearing on the call, 280° west. I walked down to the bridge and took a triangular bearing, 268° west. I knew exactly from which tree he was calling, although the animal was at least half a mile away. I had taken the same measurements before and tracked him down in the large fig tree by the Jenut Khamis saltlick. He would probably feed there for a while so there was no hurry. I headed back to the cabin to heat up some porridge and coffee.

But there was something different. I stopped in my tracks and listened again to the clear, joyful hooting, for a few moments uncertain; then there it was again, the additional notes that meant another gibbon. Stanley had found himself a mate. Suddenly his hooting ceased. There was a moment's silence, then the new female began her great-call. Three rising introductory hoots followed by a beautiful wailing scale of six longer sliding notes. Stanley joined in again with a chatter of happy twitters. Now they had really done it. Every gibbon within miles would recognize that challenge, would know that a new breeding pair was forming who would soon be competing for territorial space.

Breakfast forgotten, I hurried along the main trail into the tall, shady forest. Silver-barked merebau trees and feathery-flowered pelongs linked boughs to form a leafy ceiling, but I had no time to admire their beauty. I cut left down the Jenut Khamis path. A troop of long-tailed macaque monkeys parted and crashed away on either side, croaking their disapproving '*kra, kra*' at my disturbing presence. As I neared the small clearing around the saltlick, I slowed down and slipped cautiously behind the derelict tree-hide overlooking the fig tree. After all, Stanley was still very shy, it was an important day for him and I didn't want to upset the new pair with my over-eagerness.

The two gibbons are still calling. Fig tree 284 is covered in fruit ripening nicely. Once again the male gibbon stays quiet to allow his new companion to render her own solo of prolonged wails. He joins in again for a brief duet, hooting, and then the pair fall silent. They sit about five yards apart near the top of the tree, resting. The ripe fruit does not seem to tempt them, so presumably they have already eaten their fill before calling.

This was a most exciting piece of luck for us. Here was a new gibbon partnership in the process of forming. Something that had never been described before.[3] Something we had to try to watch in great detail.

I can see the resting animals well through my binoculars. Stanley I know well. He is a dark-chocolate animal with neat white socks and gloves and a complete white ring surrounding his black impish face. Sylvia, his mate, is very similarly coloured and about the same size as Stanley, but her coat has a dull look and her face ring is narrower and broken down each cheek. She has clearly visible black nipples and a small pink area visible in her sexual region. I get the impression that she is rather older than her new mate.

The two animals move away fast and silently to the north-east, swinging with enviable ease from bough to bough. As they cross the JT trail they slow down slightly to feed, picking off vine shoots as they pass, but not really stopping. Another gibbon group calls from the south-west on the far bank of the Lompat river, but Stanley and his new partner pay no attention. They are foraging for more vine shoots as they work their way steadily west.

We surprise a troop of banded leaf monkeys in the large keruing tree 666. They turn surprised black faces to me before dashing off through the tree-tops with harsh alarm calls, leaping over the main Land-Rover trail to the north. Tree

666 is gnarled and old, about one hundred and forty feet high and draped in vines and epiphytes. One keluot vine bears fruit which hang in long red trellises. They are tough fruit, their stones protected by brittle irritant hairs. These hairs help to discourage the many species of squirrel that gnaw and destroy the seeds, but primates can eat the fruit whole and pass the stones undamaged through their digestive tracts. The gibbons certainly like these fruit and Stanley and his new companion lose no time in settling down to the feast.

After twenty minutes they suddenly leave the tree. At first I fear they must have seen me but they have more serious cause for alarm. Imbo, the lone male siamang who lives in the eastern corner of the Reserve, climbs across into tree 666 and hurries up to feed on the keluot fruit. Siamangs are jet black in colour. We have often seen jealous encounters between the two species, with the larger siamangs chasing gibbon families away from prized food trees, but Imbo was rather unusual. He had lost his mate of the previous year and now lived a solitary and rather sad existence. Recently he had developed a strange relationship with the gibbon Stanley, who had also been a lone bachelor. Imbo did not tolerate Stanley feeding very close to him but he certainly enjoyed his company otherwise, and we had often seen the two animals travelling together, Stanley first and Imbo hurrying behind, anxious to keep up with his more mobile relative. The two friends sometimes called together, producing an absurd duet that must have greatly confused their gibbon and siamang neighbours. Yesterday they had chorused in just this way, but Stanley's bachelor song had obviously been recognized by a passing female gibbon and our assistant Kalang had regaled us with a vivid account of the arrival of a second gibbon. At first Stanley had attacked the newcomer, but soon he succumbed to her charms and the pair had wandered off together, leaving Imbo alone. The lone siamang had probably heard the gibbons calling this morning and had come to rejoin them, but it seemed he was no longer welcome.

Stanley and his new partner, whom we named Sylvia, climb up into a tualang tree, one of the tallest trees in the forest, its broad white trunk extending upwards two hundred feet or more. The gibbons sit close together in a nice open position.

Stanley climbs towards Sylvia but she moves shyly away. He repeats the manoeuvre but with the same result. Unperturbed he settles down again and she returns to the branch above him. They both gaze intently to the west but I have no idea what they can have seen there. Both seem slightly anxious, give faint hoots and come tumbling down from their high perch. The new pair hurry west for a few minutes then stop and scan the forest again. Branches are swaying ahead; someone is travelling very fast towards us. Three new gibbons suddenly swing purposefully into view, and hurtle after Stanley and Sylvia. Who can this be? We are in unoccupied territory. Then I recognize the male racing after Stanley; it is old Gilbert, clearly identifiable by his unusual two-toned coat colour. I am very surprised to see him here, for he has led his family more than half a mile outside their normal ranging limits to attack the new pair who have dared to call this morning.

All five gibbons weave back and forth in a blur of fast motion. Gibbons are hooting with rage, gibbons are shrieking with fear. One hotly-pursued animal leaps to the forest floor for safety before scrambling hastily back into the trees when his assailant has passed overhead. It is impossible to see who is who or which way the battle is going. The fighting gibbons retreat about a hundred yards towards the big tualang tree, then the battle drifts back towards my vantage point.

Abruptly the fighting ends and Gilbert and his family hurry back the way they had come. Sylvia travels with them but Stanley hangs well back. I follow the animals as far as the boundary of Gilbert's territory. Stanley seems to have been lost but Sylvia is still with the other gibbons. They are resting in a clump of very tall trees. After half an hour three animals, Gilbert, Gertie and their juvenile daughter, move on towards the centre of their range. Sylvia stays put. I go back to look for Stanley but cannot find him and by the time I return to Sylvia's tree there is no sign of her either. It appears that Gilbert's raid has been successful and that the new partnership has been nipped in the bud before it could constitute a serious threat on his eastern boundary.

Next day I found Sylvia travelling alone in the eastern corner of the study area. An hour later I came across the lone male siamang, also on his own, but of Stanley I found neither sign nor sound.

Another day passed and Sylvia met up with the lone siamang and travelled with him just as Stanley had done previously. Later in the morning I met Stanley, who was still alone, and he called for the first time since the fight with Gilbert. Next day he called again. Sylvia was still with the siamang. As soon as Stanley had finished his song the siamang began calling and Sylvia could not resist joining in, even giving her territorial great-calls at the peak of the duet. Gilbert's group called twice that morning in response but Stanley had the last word by calling again just before midday.

Having betrayed their positions so clearly, Stanley and Sylvia were easily able to find each other again, and the following morning, five days after their first duet together, the forest once again resounded to their wonderful song. This time they were not attacked and they could settle down to the job of getting to know each other and jointly exploring the new territory they were claiming as their own.

Days passed and early each morning the new gibbon pair called and I was able to locate them. They were still quite shy of human observers, hooting whenever they saw me and hurrying ahead, but their nervousness was short-lived and by the time I had crept close enough to see them again they appeared to have forgotten about the disturbance. They called from almost the same site each morning, a keruing tree not far from Gilbert's eastern boundary. With this ritual accomplished they usually travelled south across the main Land-Rover trail, moving fluently through the overhanging canopy back to the big fig tree where they had given their very first calls together. There were still plenty of figs here and between feeds they rested in a convenient tualang tree from whose lofty crown they could see right across the canopy and far over the forest. Possibly they could even see Gilbert's territory and if Gilbert's group ever climbed a tree of equal height, the rivals would be able to glare at one another as clearly as can neighbouring primates in the open savannahs. While they were resting in their look-out tree, Stanley and Sylvia got on better terms, grooming themselves and finally each other.

On 12 April I followed Stanley and Sylvia as usual, but I also asked Kalang to follow Gilbert's group. Gilbert had a busy day. Kalang reported that Gilbert's family had to fight off a raid into their own territory from their westward neighbours, group C. Gilbert and Gertie successfully drove the other gibbons back across their borders, but tempers were obviously running high and both Gilbert's and Stanley's groups called twice that morning.

Next day we again followed both groups. Gilbert's group called first. Stanley and Sylvia replied an hour later. As soon as the new group had finished their duet Gilbert and party rushed straight towards them, covering a distance of more than five hundred yards in just a few minutes with Kalang trailing far behind. Like avenging furies Gilbert and Gertie with their whimpering daughter Gail descended on Stanley and Sylvia, who were resting after their singing. The new pair were much closer bonded than on the previous occasion, however, and were not going to be split up so easily. For fifteen minutes battle raged. All was confusion with gibbons flying in all directions, black forms leaping, animals jumping to the ground. Suddenly Gilbert gave up and headed back for his own territory. Stanley and Sylvia followed for a while before heading back to the centre of their own fast-growing range.

Two days later Gilbert had another chase with group C on his north-west border and then travelled right across to the east to have another go at Stanley and Sylvia. The new duo again staunchly defended themselves and Gilbert was sent scuttling back to his own domain. Sylvia was particularly fierce in this fight and I had one clear view of her chasing after the retreating Gilbert who was screaming with fear. A sight to gladden the heart of any women's libber.

Gilbert was still intent on disrupting the fast-developing harmony between Sylvia and Stanley. He made further aggressive sallies on the 18th and 20th of April and even allowed Gordon, his sub-adult son, to return to the family fold to boost the numbers on his side. Even this extra weight failed to do the job, however, and Stanley and Sylvia began to get increasingly daring, ranging and calling over a wider area, and even using parts of the forest that were originally within Gilbert's range.

In compensation Gilbert began to make more use of the extreme south-west corner of his range, for group C's efforts made it quite clear that there was no give on his north-west boundary, and although there was no gibbon competition due south, that was riverine forest, not really suitable for gibbon exploitation and already supporting too many long-tailed macaque monkeys for comfort.

Within six weeks of Stanley and Sylvia's first calls together, the whole shift in the population was complete. Their own range was defined exactly as it remained until the end of our study. Gilbert's range modifications were also fixed. We witnessed no more aggressive interactions with group C and only one slight squabble with Stanley and Sylvia in June.

At the time of writing, a full three years after the completion of our Malayan studies, students working in the Krau Game

Reserve confirm that Stanley and Sylvia's boundaries are still virtually identical to those they had established during the first two months of their partnership. Only one thing has changed. Sylvia now carries a baby gibbon born in December, 1975.

The gibbon is the smallest and least closely related to man of all the apes, but already in this first species we can see several characteristics which man and gibbon share in contrast to monkeys. Not only does the gibbon's lack of a tail and upright posture make him look more man-like than any monkey, but his tight monogamous family life and territoriality have strong human parallels and are quite different from the polygamous, matrilinear societies so typical of monkey groups. We must see to what extent these linking features are shared by the other members of our ape list.

CHAPTER 2

Siamangs

MALAYA 1973

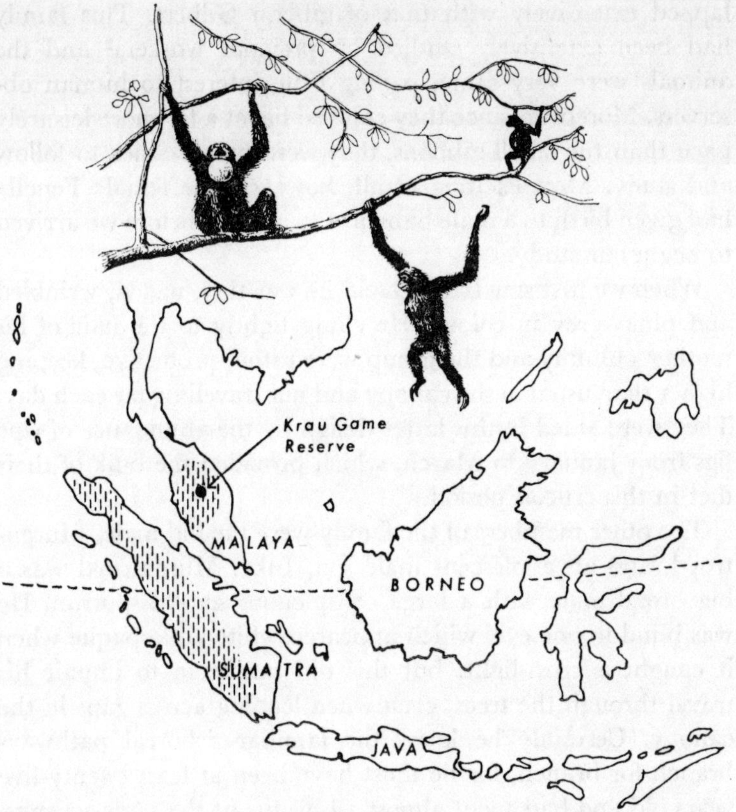

The white-handed gibbons are just one of nine different species of lesser apes distributed throughout the islands and mainland of South-East Asia. All are long-armed, territorial brachiators;[1] all are very similar in their habits. We need consider only one of these other species individually. Because of his greater bulk, about twice the size of a gibbon, the siamang stands out among the lesser apes and shows the most behavioural contrast with his smaller cousins.

Several families of jet black siamangs shared the forest at Kuala Lompat with the gibbons. One siamang family, that of Murgatroyd and Fenella, occupied a territory that overlapped extensively with that of gibbon Gilbert. This family had been extensively studied by previous workers[2] and the animals were very tame, paying little interest to human observers. Moreover, since they enjoyed life at a far more leisurely pace than the small gibbons, they were much easier to follow and study. Most exciting of all, however, the female Fenella had given birth to a male baby just two weeks before we arrived to begin our study.

When we first saw baby David he was tiny, naked, wrinkled and pinky-grey in colour. He clung tightly to the hair of his mother's tummy and the group were rather protective, keeping higher than usual in the canopy and not travelling far each day. They were aided in this latter design by the abundance of ripe figs from January to March, which provided the bulk of their diet in this crucial period.

The other members of the family were the old male, Murgatroyd, and an adolescent male son, Inky. Murgatroyd was a big broad male with a large conspicuous scrotal sporran. He was blind in one eye, which appeared white and opaque when it caught a cross-light, but this did not seem to impair his travel through the trees, even when leaping across gaps in the canopy. Certainly he knew the familiar arboreal pathways branch for branch, for he must have been at least twenty-five years old and had spent almost all his life in the same seventy-acre range. Siamangs do not even use all parts of their range. Their well-used arboreal pathways link only those trees from which they feed, or where they sleep or call.

The adolescent male was five years old; he had been just an infant when I had visited the group briefly some three years earlier, but now he was nearly as large as Murgatroyd and the developing rift was already apparent between him and his parents. He would often sit a little apart from them when the group rested and sometimes made short detours from the main family travel route. Occasionally he overstepped the mark and was chased away, or bitten, by Murgatroyd. Fenella was very short-tempered with both her quarrelsome males and if they came too close to her precious baby she would lunge at them, barking and hitting them with her wrist, or even biting them fiercely. Even so, despite these evident tensions, the whole family was remarkably cohesive.

The siamangs are sleeping in a large loh tree at the northern end of their range. It is a mile from camp so I leave early to catch them before they move off. The loh tree has a large strangling fig growing on it and it is the fruit of this fig that has attracted the siamangs to the tree several times in the last few weeks. By 6.45 the siamangs have moved from their sleeping positions but are still in the tree. The adolescent eats a few leaves, but Fenella is grooming baby David and Murgatroyd sits two feet away patiently waiting for her to move.

Quite suddenly she stands up, baby David hanging on to her hairy chest. She walks bipedally along the branch, grasping overhead supports with her hands. She swings to another bough and drops vertically into the branches of a smaller tree below. Murgatroyd and Inky follow as she leads the way north. One, two, three; one, two, three; the family pass each point in perfect line. Twenty yards north of their night tree they stop to eat the young limp leaves of a terbakah tree, but after a ten-minute snack they are off again south-west, passing quite close to their night tree again. They stop for a quick meal of menchepong leaves, rest briefly before continuing on their way. Adolescent Inky takes over the lead. The family drop gradually lower in the canopy as they transfer from one tree to the next along a well-used travel route till they reach a small tree where a keluot vine is in fruit. The animals space out around the tree crown and start eating the fruit. They are almost invisible in the green foliage but the steady crunching of hard, nutty fruits is clearly audible. The sky is still grey and cloudy and one of Gilbert's peripheralized sons is calling to the north of us. He has been calling for fifteen minutes on his own but now I can hear Stanley and Sylvia calling way off to the south-east. Suddenly Gilbert and Gertie burst into song quite close to me. Kalang is following Gilbert's family today so perhaps our paths will cross. Ten minutes later gibbon group C begin their distinctive chorus.

8.40: Apart from a sporadic nibble, the siamangs have more or less stopped eating; they rest until 10.20 and I have little to do but listen to the other groups. The single male gibbon is still calling to the north but otherwise the forest is fairly quiet again. The sun comes through. The gibbons Stanley and Sylvia call for a second time. Two rhinoceros hornbills fly overhead, honking harshly. Then a group of siamangs south of the river start calling with loud, reverberating calls. They sound so close that I find it difficult to believe that they are nearly a mile away through the dense forest.

Murgatroyd and Fenella do not respond to the challenge from across the river. They partake of more fruits until 11.00, then move on south-west to another terbakah tree. They rest for an hour before indulging in a forty-minute meal of the soft, new leaves. They are big leaves, about a foot long, and with the thick texture of young cabbage. They are rather soapy to human taste but not bitter like most of the forest leaves. They are certainly a favourite with the siamang and constitute a large portion of the animal's spring diet. The siamangs halt for a few minutes whilst light rain falls, then move into a neighbouring tree to vary their leafy meal with vine shoots. A large troop of dusky leaf monkeys travel past, eating as they go and with no sign of hurry. They turn to stare at me, their strange white-ringed eyes and white muzzles giving them a spooky, clown-like appearance. One of the females has a bright orange baby. The colour marks it out as special: a delicate and valuable member of the family in need of extra care and consideration. The signal is as clear to the other monkeys as it is to me, yet to a predatory, colour-blind carnivore this same infant would be the least conspicuous animal of the whole troop.

By 13.00 the siamangs are travelling again, halting only briefly then flowing on with Fenella back in the lead, little David clinging tightly to her side, and the two males following close behind. They arrive at a large fig tree and start feeding vigorously on the ripe fruit. Murgatroyd pumps up his throat pouch and lunges angrily towards Inky when the latter comes too close, then he resumes feeding again.

At 14.00 the gibbons arrive. Gilbert leads the way, followed by Gertie and Gail. The sub-adult male, Gordon, is a little way behind, part of the family again so long as his presence is needed as an extra check on the territory-grabbing Stanley and Sylvia in the east. The gibbons look hopefully at the fig tree. Gilbert crosses into a high corner but Murgatroyd rushes to chase him out. The gibbons leave as quickly as they came, and I catch a glimpse of Kalang as he hurries after his restless quarry. The siamangs feed for another half-hour and even infant David, now five months old, tackles a few of the soft red figs.

At 14.30 the family are off again, travelling in a wide arc to the south. They arrive at a large terbakah tree and start to feed. It is their third meal of terbakah leaves that day and it is also their largest, for they are still feeding there an hour and a half later. Again Inky comes too close to the feeding Murgatroyd who lunges after him, pumping up his throat pouch. Inky squeals and flees, then climbs higher up to the top of the tree to feed. Murgatroyd feeds longer than the other members of the family who wait patiently for him. At last he has eaten his fill and the group sit quietly, scanning the surrounding trees. The dusky leaf monkey troop is just up the hill, feeding on new merebau and kempas leaves.

When the dusky leaf monkeys move off, the siamangs climb into the merebau tree themselves. They have finished their foraging for the day and move to their familiar sleeping positions in the tree. They have over thirty different night trees in their range and use the most convenient each evening, wherever their range may have taken them. This is one of their favourite night trees and every time they use it they adopt the same personal sleeping positions.

By 17.00 my job for the day is done and I head back for camp. Kalang will probably be there before me, for the gibbons are the first primates to clock off each day. About three-quarters of an hour after the siamangs have retired, the long-tailed macaques and dusky leaf monkeys will be settling down for the night and last of all, just as dusk is falling, the banded leaf monkeys will move into their own regular night trees. Nightjars circle over the camp clearing, giving their plaintive calls as they hawk for high-flying insects in the fading light of the setting sun.

It had been very much a typical siamang day. An easy-going day. Half a mile of travel, five and a half hours' feeding. Very little social interaction within the family, no interaction with neighbouring groups of siamangs and only a slight brush with the more active gibbons. Sometimes the routine is varied with a bout of territorial calling. Or on very hot days the siamangs might visit tree-holes to scoop up and drink the black mosquito-ridden water which has collected there. For a change in diet the siamangs may visit one of their insect trees to gain some protein by licking up termites from the winding columns which scurry up and down the trunks. Mostly, however, the siamang diet is a mixture of leaves and fruit and I could now add another day's activity and carefully-mapped travel route to our fast-growing pile of data.

Plotting out the actual travel routes of the siamangs was very revealing. Siamangs actually spend more time feeding per time spent travelling than any other primate studied in the wild. Since they feed largely on fruit which is a rare food type in the forest this implies extraordinary efficiency at finding food, and we were keen to see how they achieved this. Each month Murgatroyd's group and the lone male siamang, Imbo, were followed for at least five consecutive days and their routes were plotted precisely.

During any five-day period there will be one, or sometimes two, favourite trees which are at the cross-over points of loop-shaped travel routes, and these are the most exploited trees over this period. The best tree will be visited nearly every day and often it will be the first food tree visited. During the rest of the day the siamang visit several other food sources to complement and balance their diet. The hot tree of the moment is usually a fruiting fig tree.

Over a few days the siamangs use very efficiently a small portion of their range around the hot tree, then every few days they set off on a wider loop to some other part of the range. Possibly these safaris are to patrol the security of the range, making sure there are no trespassers and checking up on the whereabouts of their neighbours, but I think the most important function of these forays is to check out oncoming food supplies in other parts of their range.

Murgatroyd's family's use of fig tree L.25 demonstrates very well the degree of planning and spatial calculation that must be used in siamang route selection. During January, February, March and April the siamangs had lived almost exclusively in the western half of their range, but on 3 May they made a major excursion to the eastern end. They visited tree L.25 which was just developing small green figs. The siamangs ate a few green figs but returned the following day to the west. They continued to feed at several good food sites, but five days later headed east again. They travelled to within sight of tree L.25, presumably saw that the figs were still not ready (no red fruit visible, no activity of squirrels, monkeys or hornbills) and returned west again for another three days. On 12 May they returned to L.25 just before dusk. Next morning they fed in L.25 but the figs were still green and the siamangs returned west. By mid-June they were visiting L.25 approximately every other day to crop off the few new figs that ripened each day. Two-day intervals allowed sufficient fruit to ripen to make a visit to the tree profitable. By the end of June the figs were ripening faster and the siamangs visited the tree every day. As the fruit started to become scarce again in July the siamangs reverted to a system of visiting the fig tree every second evening for the last meal of the day, and a snack the next morning. To plan such routes efficiently the siamangs must know the rate that figs are ripening in the tree, the food needs of the whole family and consequently how long it will be before the tree is worth revisiting. Then they must plan a loop-shaped route to take in the best alternative food sites in the vicinity while more fruit is ripening. Every now and again they take time off to make a recce into more distant parts of the range just to see where the next good fruit crop is going to happen. Sometimes the siamang may be attracted to good food trees by clues such as concentrated monkey or hornbill activity, but normally the siamang is the first to discover each new food source and his high intelligence, long memory and very intimate knowledge of his limited network of paths through his small home range probably account for the animal's great feeding success.

The family cohesiveness and co-ordination of the siamangs was in marked contrast to the behaviour of the gibbons, who spread out far more when they fed, sometimes eating in different trees. In siamangs the distinction between travel and feeding was clear. The group travelled as a unit from one food tree to another, then stopped and had a good solid meal. Gibbons, on the other hand, fed casually hither and thither as they travelled, often snatching as they passed without really stopping at all.

The siamangs' day started about forty minutes later than that of their gibbon cousins and it ended later in the evening. Their calling time was also that little bit later. It seemed important for the two species to have their own slightly different calling times, so that they could call and hear their conspecific groups' whereabouts without obliterating the other species' calls with their own. Siamangs called less often than gibbons, only once every three days on average, and rarely more than once in a day. What they lacked in frequency they made up for in both vigour and volume. If the chimpanzee is regarded as the loudest animal in the African forest, so the siamang is undoubtedly the noisiest in Asia. The resting siamang group suddenly bursts with urgency, the black apes clamber higher and higher in the canopy to a platform in a tall open-crowned emergent, such as a tualang or merebau tree, and they begin their rhythmical and explosive series of barks. Both male and female siamang have strange spherical throat pouches which inflate like large pink balloons under the chin. As the pouch rises and falls it emits an eerie hollow booming sound which, though not very loud, carries a fair distance, and can easily be heard from a mile away. Such territorial calls are a display of family strength and unity. It really matters to be heard at maximum power. The gibbon has less need of a throat pouch because he is smaller. He has more time on his hands and travel is less expensive. He can afford to call four times as frequently as the siamang and can visit and call from all his boundaries, even making raids deep into the neighbouring terrain to reinforce his territorial claims with physical chasing and fighting. The siamang cannot afford such excesses. He has twice the bulk to support and must spend a longer time feeding each day. He must choose his travel routes more economically than the more affluent gibbons and rely almost entirely on the long-distance effects of his calling which he reinforces with the pouch boom and visual display.

When Murgatroyd's group called it was the adolescent male, Inky, who shouted most vigorously. Often he moved to an adjacent tree to his parents and called directly at them, rather than with them, prolonging his calls long after they had given up. This was just another sign of the growing tension between parents and son Inky, a tension that would eventually lead to his leaving the family.

Siamangs may not call very much normally, but quite by accident we discovered that they could greatly increase their territorial efforts if they felt socially threatened.[3] One day when we had been tape-recording a call bout, we played back the call to listen to it. Suddenly three horrified black faces peered down at us and the siamangs shot off like gibbons, gave two more calling bouts themselves that day and travelled twice as far round their territory as their seasonal norm. We repeated the experiment a few days later with the same result.

We were fascinated by the siamangs' interactions with their near relatives, the gibbons. Both species seemed to be living in the same area and exploiting so many of the same resources. How could one species survive the competition of the other?[4]

Siamang ranges were only half as large as and entirely overlapped by those of the gibbons, but the siamangs' more concentrated activity was in exactly those parts of the gibbons' territories where the gibbons also spent most of their time and obtained most of their food.

There were slight differences in diet – siamangs ate rather more leaves and insects than did the gibbons, but the two species did share about eighty per cent of the items in their diet and obtained a high proportion of their food from the very same trees.

When siamangs were present in a tree, the smaller gibbons rarely got a chance to feed. Murgatroyd would bark at the gibbons and chase them away and even Imbo, who sometimes ranged with lone gibbons, was intolerant of them if they came too close when he was feeding. Siamangs could obviously monopolize the best fruit trees whilst they were there, but they could not stay all day, and the gibbons had plenty of opportunity to feed when their larger relatives were elsewhere. Indeed, the earlier daily starts of the gibbons meant they had often fed and left these trees before the siamangs arrived.

The really crucial difference between the two species seemed to lie in the different travel costs per food reward that apply to their different sizes. Siamangs travelled an average of just over half a mile a day and fed at only eight feeding sites a day, whilst gibbons travelled more than twice as far and fed at over sixteen sites. In an average five-day period, the siamang family obtained fifty per cent of its diet from only three trees, whilst gibbon families visited ten trees for the same proportion. Gibbons could make a living from travelling to small or dispersed sources of food that siamangs could not afford to visit. The gibbons were more efficient at exploiting food sources, but so long as siamangs could behaviourally dominate gibbons at the best sites they could not be outcompeted, and the two species could live together.

By the end of our study the infant David was nearly eight months old. We had watched his progress from an ugly, wrinkled, almost naked and very wobbly creature to a furry active bundle of black fun. He still rode about on his mother, slept with her at night and suckled from her, but he was now sampling the occasional fig or succulent vine shoot for himself. Often he wandered several feet from his parents to gambol on his own, twisting about, slowly swinging on vines, gaining new confidence and skill every day. He was showing increasing interest in his father and when Murgatroyd and Fenella rested or groomed together at midday, David clambered about on his daddy's head, biting him playfully. Far from being annoyed, Murgatroyd seemed to love it, gently poking the infant, tolerantly permitting David to take uncomfortable liberties with his person, and sometimes getting so excited himself that his throat pouch blew up like a great pink balloon. We were fascinated by this developing paternalism, for in a few months a remarkable change would occur within the family, a change almost unique to the siamangs. During the second year of its life the infant siamang transfers from being carried by the mother to riding on its father.[5] When I had seen the group three years earlier Inky had been a second-year infant and Murgatroyd was responsible for his carriage. In Sumatra, too, I had seen an infant carried by his father. Such babies still sleep with their mothers at night, but by relieving the mother of this extra load the male siamang gives her a chance to build up her strength again for the task of starting a new baby, and we must assume that this either enables her to breed again sooner or at least gives the next baby a better chance in life.

And Inky? What would befall him when he finally broke away from the parental family? This, too, we could probably answer. When I had seen the group in 1970 Inky had had two older brothers, Sammy and Julian. By 1971 the eldest had left the family, attracted two females and set up a new group to the east of Murgatroyd. Sammy had failed to keep control. One of the two new females remained, but with a newcomer male, Imbo. Second son Julian had left the parent family and now he and brother Sammy were living together outside the parents' range to the west. The departure of siamang sub-adults is very similar to that of gibbons; they leave the family but continue to live adjacent to, but outside, the parental territory. Gibbons' larger territories adjoin, so there is virtually no free land between them, but the small siamang territories cover the forest less completely. There are plenty of gaps where young hopefuls can sit and wait, trying to attract a suitable mate with whom to set up a new breeding territory.

The siamang, like the other smaller gibbons, is monogamous and territorial, something absolutely characteristic of the lesser apes and quite distinct from anything seen in the Old

World monkeys. Why should this be so? Why do they not live in larger social groupings like monkeys?

Basically, the answer lies in their dietary specializations as fruit eaters. Fruit is a scarce resource in the forest, especially when compared with the abundance of leaves and shoots that form the mainstay of monkey diets. Monkeys, particularly long-tailed macaques, will and do eat large quantities of fruit when they find them as opportunists, but they are not so dependent on these irregular and dispersed windfalls. The lesser apes have a greater brain size than monkeys, and their higher intelligence enables them to specialize in locating soft fruits efficiently. If the lesser apes tried to shift their dietary preferences too far towards the commoner, leafy foods, they would run into direct competition with the numerous monkeys. Our Malayan study area which supported only twelve lesser apes provided a home for a total of one hundred and forty-eight monkeys of three species. Competition against such numbers would be far too fierce. The lesser apes can do something that the monkeys cannot do and they do better to stick to their specialized way of life. Granted that they must depend heavily on a scarce food supply, we can see how severe is the competition for that food within the lesser apes. Even at group sizes of four, the Gilbert gibbons and Murgatroyd siamangs had to travel further each day and still accept a slightly lower fruit proportion than their smaller-grouped neighbours, the Stanley gibbons and Imbo, the siamang. Every additional mouth to be fed means more food sources must be found and more travel to find them. The lesser apes have opted for the smallest permanent breeding group possible, the single pair, and the jealous defence of whatever space is necessary to support them.

On the Mentawi Islands, off the Sumatran coast, lives the Kloss' gibbon.[6] There are no long-tailed macaques on the islands, with the result that there is less competition for fruit and the gibbons can live at higher densities than anywhere else. In theory they could have increased their family sizes within gibbon-sized territories. Increased group size would not severely tax the foraging resources of the animals and it would theoretically be possible for members to enjoy more than one breeding partner. Male Kloss' gibbons would love to have two wives; female Kloss' gibbons would probably love to have two husbands. In fact they have retained the gibbon formula of strict gibbon monogamy and live in tiny territories of only twenty acres or so. The consequences are intriguing and amusingly human for in Kloss' gibbon males and females sing quite separate territorial songs and only fight members of their own sex. Males make sure that no other males intrude within their territories whilst the females keep out any female competition. Monogamy is preserved only by high sexual jealousy and social communal life has never been realized.

In addition to monogamous territoriality, the siamang exhibits a high level of paternalism – another human trait not seen in the Old World monkeys. Male baboons and Barbary macaques sometimes form relationships with particular infants,[7] but these are rarely their own offspring and are really a ploy of social buffering, whereby the male is protected or even raised in status because of the presence of his infant. What the siamang shows is something quite different: a true relationship between a father and his infant within the context of a tight-knit family group. It is certainly much more like the paternal behaviour we see in human families, but it would be premature to derive any theories of the common derivation of such behaviour without first examining our other close relatives, the great apes.

CHAPTER 3

Gorillas

ZAIRE 1975

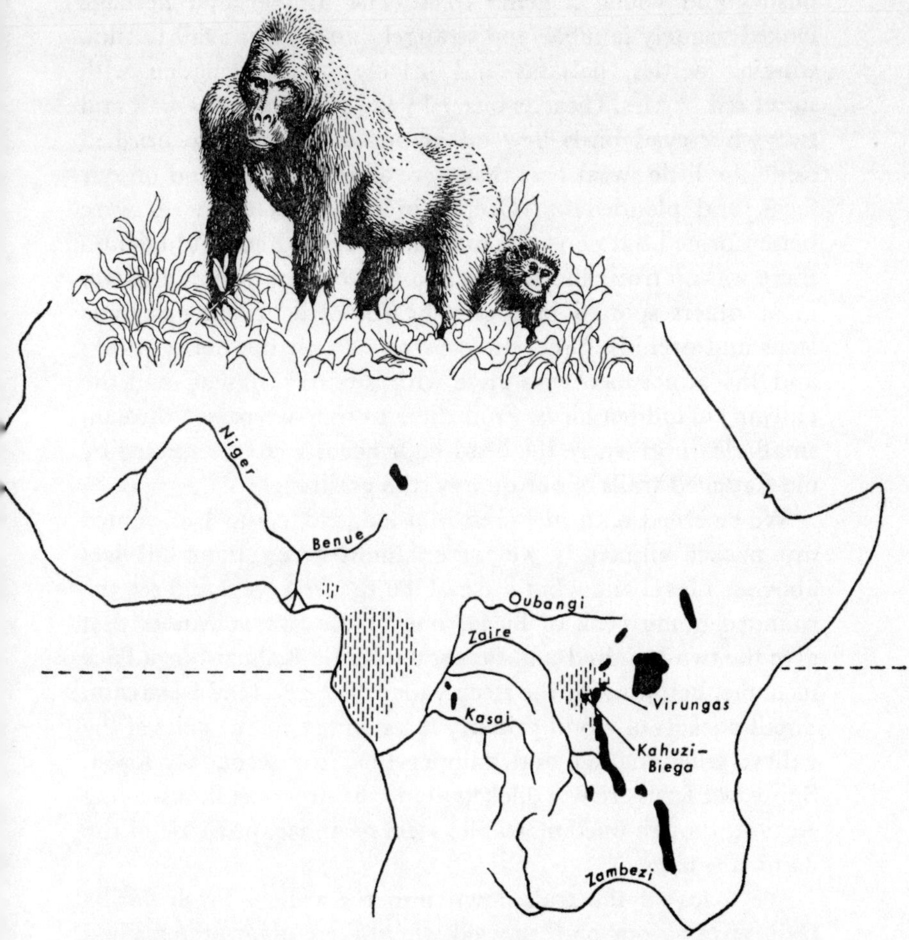

We picked our way up the slope between the glossy green tea bushes which prosper on the rich, red African soil. Below us lay

Lake Kivu, shimmering mysteriously in the morning mist, surrounded by a lazy landscape of sensual, rolling hills. Above the plantations we paused for breath before plunging into the dense thickets of secondary forest. My pygmy guides led the way along the rough path that twisted among the tall musanga bushes and young hagenia trees. The lush ground herbage looked vaguely familiar and strangely un-African: tall thistles, stinging nettles, balsams and prickly brambles laden with sweet red berries. Cicadas buzzed with a monotonous wail and two white-eyed birds flew off at our approach. We brushed aside the little sweat bees that were attempting to land on our faces, and plodded on. Deeper in the forest the trees were better formed but none was more than a hundred feet high and there was no true canopy layer. Some were covered in lush wet moss, others sported long grey beards of lichen or epiphytic ferns and orchids. Creepers grew thickly on the horny trunks and the atmosphere was alive with sounds of insects and the chirping of hidden birds. From time to time we passed through small clearings where the head-high herbs were intersected by old flattened trails of our quarry, the gorillas.

We emerged on to the crest of a long ridge and I examined my pocket altimeter; we were almost seven thousand feet above sea level and what a view! To the west we could see the rounded dome peak of Biega, one of the two volcanoes that give the two-hundred-and-fifty-square-mile Kahuzi-Biega Park its name. Below us lay the steep valley of the Karembiri stream. Small pockets of virgin primary forest clung to the sides of the valley, taller, darker and damper than the secondary forest. Spiny tree ferns grew in thickets along the stream gulleys making elegant, though uncomfortable, scenery under the shade of the dominant trees.

We followed the trail down into the valley. Fresh gorilla trails crossing our path showed where these great animals had bulldozed through the undergrowth in their systematic foraging. One trail intrigued us, for it had not been there the previous day and was obviously made by a lone animal living apart from the group that we were approaching. We reached the valley bottom and crossed an open marshy clearing, stepping carefully between the reedy tussocks. It was along this stream bed I had first seen the gorillas two days earlier. Now their tell-tale tracks showed us where they had moved on.

We examined the little bowers where the great apes had taken refuge from a midday rainstorm, snug little nests sheltering among the vines at the base of trees and under leafy bushes. Everywhere the vegetation was flattened by gorilla activity and clouds of tiny flies clustered on the crushed growth, sipping at the leaking sap. The shiny three-lobed dungs told another story of who had been eating what, and stripped vine stems and broken twigs lay scattered where the gorillas had made their meals.

We followed the wide trail where the silverback male, Mushamuka, had led his group away in the evening. They had moved off in a leisurely manner, foraging as they went, but the big male's direction had been sure and purposeful. He knew exactly where they were going, exactly where they would make their nests for the night and exactly where they would feed the following morning, although the group had been absent from this part of their range for several months.

The nesting area was situated on a short detour from the gorillas' main travel route. There were twenty-one animals in Mushamuka's group, but since infants sleep with their mothers we found only sixteen nests. Five of these were up in the trees, built in supporting limb forks and constructed from bent and twisted branches neatly bound together to make springy, cup-like structures. The larger animals had bedded on the ground, leaving rounded depressions in their less elaborate piles of herbs. Most of the beds were soiled by still-moist gorilla dung. I was interested in this apparent lack of hygiene. Other apes are very careful to avoid fouling their nests or their fur, but the diet of the gorillas is so fibrous and the dungs so dry that these do not soil the fur even when lain on and the gorilla's need of hygienic behaviour is far smaller than in apes with a fruitier diet.

The gorillas had climbed the steep hill; their individual foraging trails plaited across each other through the six-foot nettles. Broader trampled areas showed where the great apes had paused to feed. My guides became more cautious, stopping and listening, then creeping on. Suddenly the peace was disrupted by violent roars ahead of us, followed by a rapid '*tok-tok-tok*' as a large male gorilla beat his chest in threat. The gorillas were just ahead, over the next shoulder of the hill. We continued upwards to emerge above them at a point from where we should have a good view of their activities but be in little danger if the silverback male charged us. As the guides flattened the herbage, I crept silently forward to peer over the ridge.

The gorilla group was straggled out below us. Mushamuka, the great silverback leader, lay in a clearing, glancing at us suspiciously but not sufficiently worried to raise his four-hundred-and-fifty-pound frame from its resting position. A few feet away a female carrying a tiny black baby had uprooted a small bush and was carefully divesting the twigs of their bark before transporting them leisurely into her gaping mouth. The rest of the group was scattered over about seventy yards. Movements among the branches and herbs showed where animals were still feeding or juveniles were playing. We could hear occasional rustlings, twigs snapping, coarse scratching noises and even Mushamuka passing wind. Down the hill to our left an animal repeatedly beat its chest but the displayer was hidden from our view by the trees. The group was resting after its substantial morning forage. The animals would remain relatively sedentary for the next few hours.

Two baby gorillas climbed up a willowy tree just twenty-five yards away. One infant sat in a fork swaying the tree gently to and fro, clapping his little hands and scrutinizing me with bright button eyes. The other thumped her tummy with one fist then, to improve her infantile display, picked a bunch of leaves and, with them clenched between her teeth, continued her bizarre performance.

The tree was obviously not going to take the weight of both animals and the second sensibly descended. Perched in a crotch about twenty-five feet above the ground the remaining infant swung the tree about. His playful girl-friend climbed up into a bush immediately behind him. She was slightly larger and fluffier with a reddish tinge to her dark coat. She stood upright in the top of the bush, beat her chest then flopped down, a shaggy black bundle. Her playmate still swayed backwards and forwards on his slender mast. He hung by one arm and one leg, smacking his hairy chest with the free arm, then somersaulted over with casual ease. Another, even smaller, baby climbed up and joined the reddish female in a game of thumping and tumbling. The larger infant stood upright and her new companion came up behind, placing an arm on her back to form a tandem. The female lunged forward and grabbed the black male as he swung past on his sapling. She dragged him off into the top of the bush and all three youngsters somersaulted and grappled delightfully in front of us.

Mushamuka rose purposefully to his feet. He turned his calm, blemishless face towards us, then headed off down the slope into the trees. The incredible breadth of his great silver-haired back rocked solidly with every step. He was a truly magnificent gorilla. The female with the baby followed after him, the infant clinging on to her back like a tiny jockey on a gargantuan mount. Just short of the trees the mother stopped and lay watching us with a disinterested gaze. The infant was fascinated, however, staring at us with thumb in mouth. A young juvenile gorilla climbed slowly into view up a slender muvula tree and, staring playfully at us, thumped his little chest in mock threat. Climbing higher, he began twisting and cupping the branches to make a play-nest.

It was wonderful to see the gorillas at such close quarters and so unbothered by our presence. Some animals are an anticlimax when you finally see them in the wild: they look no more exciting than in a park or zoo; but not the gorilla. The gorilla is a truly impressive animal with a very real and powerful presence. Normally gorillas are shy of man and would not permit such close inspection. The fearlessness of Mushamuka's group was a tribute to several years of courageous perseverance by a one-time Belgian tea-planter. Adrien Deschryver had visited the two gorilla groups residing in the forest beyond his plantation for six years and by his quiet, calm and unaggressive manner he gradually gained their acceptance. Getting the gorillas to transfer their trust of him to other humans was another problem, but eventually the Kahuzi-Biega became a National Park with African staff trained to take visitors or scientists out to see the gorillas. Tolerated as we were, however, gorillas are not animals with which to fool around. I could see that my guides were frightened, extremely cautious in their every movement. They did not enjoy our proximity to the great apes and came into the forest only because they were paid well to do so. Their fear was not without foundation, for when forest pygmies still hunted gorillas with bows and spears many of them died at the ferocious hands and jaws of the enraged apes. Adrien Deschryver once found the remains of a pygmy who had been torn apart by a gorilla and strewn about the bushes. There is a widespread belief that if you stand and face a charging gorilla you will be quite safe. Curiously enough for Europeans this maxim seems to hold. European scientists like Deschryver have found that the gorilla will not attack a person who stands his ground, but the image of the gentle giant is as misleading as the older idea of the savage killer. To make a threatening move in the proximity of wild gorillas is a sure way to provoke a charge, and to then show fear by running away could invite a real attack.

It began to rain, just a drizzle, but the gorillas moved into the shelter of thick bushes and there was little sign of activity. We crouched among the broken herbage and I drew my cape over my camera. The rain continued for twenty minutes or so before the clouds parted and thin streamers of sunshine pierced the gloom on the other side of the valley.

After twenty more minutes the clouds drifted away over the ridge and the whole valley was bathed in sunshine. Wisps of steam rose from all sides. A gorgeously-coloured bee-eater flitted among the branches above us, launching down to snatch insects with a loud click of his beak then returning to his perch to eat his prizes at leisure. Clouds of tiny flies rematerialized and a gaudy Charaxes butterfly hovered with stiff wings before settling on a fresh gorilla dung to probe this apparent delicacy with its long proboscis.

There was no sign of the gorillas and I was afraid they had moved off down the hill unseen and unheard in the rain. We decided to follow their trail and set off cautiously down the slope. My tiny guide, Patris, led the way and I helped him flatten the herbs in front of us while the other guide hung back to watch and listen for any tell-tale movements ahead. The theory behind this method of progression is that although gorillas will charge right up to humans if they have vegetation around them, they seem shy of charging out into the open. We reached the bushes where the gorillas had feasted earlier, and followed their narrow track through a gap between two dense thickets. Patris and I had just entered a tiny clearing where the herbage had been crushed by lazing gorillas when there was a grunt ahead and our lookout behind us yelled '*Ngila! Ngila!*' (gorilla).

> Terrific barks, followed by a crashing charge, and Mushamuka is screaming at us from only a few yards distant, hidden by the bushes. We can see nothing and Patris tries to run back along our path, but I grab him and hold him as it would be dangerous to pass through the thicket so close to the angry silverback. Patris struggles but I refuse to let go as the roaring gorilla charges towards us. He is huge, a

mountain of powerful muscle. We stand facing him as he dashes to within four yards and then stops in the clearing, screaming, tearing up herbs with his enormous arms and hurling them into the air. My heart pounds and I am dizzy with fear, but remain absolutely still, gazing into his eyes. I can see the terrible scarlet lining to his yawning mouth and his great white teeth; his powerful, musty odour fills the small clearing. Suddenly he relaxes, closes his mouth and looks completely calm. He turns and ambles back into the thicket, and we creep silently back along the path to the other guide, shaken by our experience. We had been wrong in thinking the gorillas had left. Obviously they had been resting in the thicket and were most disturbed by our rude interruption.

Patris was furious at my behaviour. In front of his sympathetic companion he mimed a graphic repeat performance of Mushamuka's charge and how I had prevented him from running away. He was sure I had wanted to make an offering of him to the angry gorilla and it was several days before I could persuade him that I had acted in the interest of our common safety.

We could hear that the gorillas really were moving down the slope now from the occasional breaking of branches and rustling of leaves as youngsters played and adults fed, but we could see nothing. We followed behind the group more cautiously after our recent encounter. A large female gorilla moved into an open space and fed solidly on the vine leaves that surrounded her. A small black baby clambered across her bare grey breasts and round on to his mother's back. Mother slid her hand along the vine stems drawing off all the leaves, then popped them into her frantically busy, chomping mouth.

Excited gorilla screams told us the group was moving off again. The female and baby vanished silently back into the vegetation, small tree crowns swayed as heavy apes transferred from one tree to another. The animals were moving south round the shoulder of the valley. It was difficult for us to follow so we moved round and took up positions on the opposite slope, anticipating their approach. We waited in vain for a quarter of an hour, then heard gorilla screams further down the slope. They had rounded the ridge, but too far below us to be visible. We moved after them but they were heading into very dense forest. We would not get any more good views and decided to leave them to it. We could always pick up their trail again next morning to reconstruct their evening's activity.

We climbed back up the steep side of Karembiri valley and were nearly at the top when we heard gorilla roars just below us in a deep, wooded gulley. A large gorilla rushed across the canyon, glaring at us briefly before moving away into a thicket of tall nettles. I wanted to go nearer but the guides would not go down. It was Musharamina they said, a lone gorilla who sometimes hung about the other groups looking for stray females, but they were afraid to follow a lone male. His presence explained the fresh gorilla track we had crossed in the morning and it added to my own understanding of gorilla ways. We headed back towards the main ridge as the mists descended. The fog grew thicker and thicker and my guides became less and less sure of their way. We were lost in a maze of old gorilla tracks and could see only a few yards ahead of us. The guides wanted to wait for the fog to clear, but after consulting my compass and my notes I was confident of the right direction and led the way off the mountain slopes. At last we reached a path that the guides recognized and we hurried back down to the tea plantations.

The Karembiri valley formed the southern ranging limits of Mushamuka's group. Over the next eight days they moved about four miles north through the secondary forest bordering the edge of the reserve, towards the main Hombo road. I watched them feeding on wild banana trees and the knobbly, unripe fruits of the common secondary forest tree mbuamba. The ground beneath one of the feeding sites was strewn with the debris of a paper-wasp's nest which had been raided for grubs by the gorillas. Where the main Bukavu-Hombo road cut through the Kahuzi-Biega reserve the range of Mushamuka's group overlapped that of a neighbouring troop led by another large silverback male, Dieter. This group had also been befriended by the warden, Adrien Deschryver, but Dieter was a younger, stronger male who had only recently taken over the group from the previous patriarch, Kasimir. Dieter had arrived to challenge Kasimir and for a period the two males had indulged in fierce displays and charges, but then Kasimir had fallen ill from an infected wound and had died, leaving the way clear for young Dieter.

I spent a few days watching Dieter and his entourage, but they were ranging so far away from the park headquarters that it took nearly three hours to reach them each morning and we were only able to see the animals for about four hours a day. Most of the group members were extremely tame and we were able to approach within a few yards of some of the females and young males. Dieter himself, however, was still very wary of humans and although he charged us repeatedly we rarely caught any glimpse of him, as he would not emerge from the protective screen of dense thickets. Dieter's group consisted of just eleven animals. Only a year ago it had numbered nineteen, but several members had left the new management. Musharamina, the lone blackback male we had met in the forest, had been a member of Kasimir's group, but was not prepared to stay with Dieter. It was Musharamina who had lured away several of the group's females, only to lose them again to Mushamuka. So Mushamuka's group had grown, Dieter's had shrunk, and Musharamina had become a solitary loner. There were still a few juveniles in Dieter's group but no infants. I was very happy to watch his troop, however, because they were feeding in undisturbed primary forest and it was an enormous pleasure to see how arboreal these greatest of the great apes could be, climbing and feeding in trees, nesting and swinging

Tarzan-fashion among the lianas. The forest was dark, the terrain steep and rocky, gorillas clambered up spiny tree ferns to crunch noisily on the tender growing shoots. Most spectacular of all, a large female and her juvenile were feeding on a mistletoe-like growth in a high emergent some eighty feet above the ground. We moved close up under the tree so that the descending animals had to pass within a few yards of us when they slid down the tall trunk. Even the larger males climbed trees and although we did not see Dieter up aloft, his predecessor Kasimir was renowned for his prowess in harvesting the sweet mbuamba fruits in the canopy.

Both gorilla groups used their large ranges very economically. Every day they foraged through a different area, so that they were always surrounded by fresh growth. They never stripped any area completely, so that should they need to recross their trail in the near future there would still be some food available. Normally, however, a feeding area was not revisited for weeks, or maybe months, until it had recovered from earlier exploitation and there was a good flush of new growth.

Different types of food are available only in certain parts of the gorillas' ranges and the animals' seasonal movements take into account the local and temporal availability of food types.[1] For instance, the mountain bamboo grows only on the higher slopes and shoots will only grow fast when there is heavy rain. These shoots are a nutritious and abundant food source when they are available, so during the winter months when the rainfall is greatest, both the gorilla groups move up into the higher ground to feast on the bamboo. While the smaller animals sample the tender young growth the big males crack open the hard cane nodes to eat the rich food stores protected inside. Although the gorillas cover considerable distances in a few days, in any one month their activities are centred round a relatively small part of their total range. In the course of a year, however, each group works its way round its entire range. This policy of cropping an area then moving on to allow the vegetation to recover suggests that the gorillas are expert conservationists. I am sure that we could learn much from this great ape's example of how to get the most out of an environment without over-exploiting and destroying its resources.

Although most wild gorillas both in West and Central Africa live in tall rainforest as rich as or even richer than that in the Kahuzi-Biega Park, most of what has been written about gorillas[2,3] relates to the small population that lives at much higher altitude in the Virunga volcanoes in poor montane forest, and even living above the treeline on the cold alpine mountain slopes. I camped for a few nights up on the Virungas just to see how different the habitat really was from the forests of Kahuzi-Biega, and was very struck by the breath-taking beauty of the mountain scenery and by the chilly night temperatures which left me with a terrible headache each morning.

In Kahuzi-Biega the gorillas lived in the forests below the bamboo zone, but in the Virungas they lived in the forests above it. These montane forests were composed almost entirely of only two species of tree and few trees were more than fifty feet high, whilst in Kahuzi-Biega the forest was much denser and composed of about twenty-six different types of tree, with many specimens over one hundred feet high. Far more light could penetrate the Virunga forest canopy to support a lusher herb layer of nettles, thistles, wild celery, docks, brambles and bedstraw clingers. And through the herb layer a maze of gorilla trails showed me where the animals had wandered, where they had fed and where they had nested.

Compared with the Mount Kahuzi gorillas, the Virunga population are more terrestrial in their habits – rarely nesting in trees and finding most of their food on the ground. They are unable to find such a wide variety of food types and are not able to obtain so much fruit. Nevertheless, the herb layer in the Virungas is thicker and more succulent than on Mount Kahuzi and gorilla group ranges there are smaller, with overall density of animals higher. The Virunga population must face much colder night conditions, for it often freezes at these great heights, and they are consequently much hairier and shaggier-looking than the Kahuzi gorillas.

These differences are all minor, however. It is in the similarities of the two populations that we must search for the basic nature of the gorilla, what makes him distinct from the other apes.

In both areas gorillas live in groups (up to thirty animals) of relatively stable composition, occupying large home ranges (five to thirty square miles). The ranges of different groups overlap extensively, but each group maintains fairly exclusive use of its own central area. Some lone males live apart from their parent groups. Each group is led by a patriarch, the largest silverback male. This animal plays the major role in determining travel routes, sleeping areas and the timing of group activities. He is the animal who plays the main defensive role whenever the group is threatened by man, predator or other gorillas. The silverback is the central social figure in the group and the only link binding together the different female family units. In monkey troops adult females develop close social relationships among themselves, they travel close together, sit and rest beside each other and groom each other regularly, but in gorilla groups each female remains very much independent and there is little social grooming. Over ninety per cent of all the group's vocalizations are emitted by the patriarch. When a patriarch dies or becomes weak, his leadership usually falls to one of the peripheral lone males who has been living apart from the parent group for a few years, rather than the second-ranking male. There is constant competition between overlapping groups for females, and these are often kidnapped back and forth in marked contrast to monkey troops, where it is males who move from one troop to another.

In both areas troop ranges include several different biotypes or habitats, and gorillas use these in a clever and conservationist manner so as to obtain the best seasonal range of food without ever over-exploiting a particular area. Gorillas are almost entirely vegetarian browsers, munching their way through the forest in much the same way as a herd of buffalo. To support their great bulk they must eat large quantities of common plants. Fruit or particularly succulent vines and epiphytes may tempt the big apes to climb trees to feed, but at such great size arborealism is a risky way of life and animals prefer to travel along the ground, feeding on low vegetation.

In the Kahuzi region gorillas show preference for rather open secondary forest and in the Virungas they show preference for the open herbaceous slopes. To a certain extent these apes can be thought of as primitive gardeners, for their own activities undoubtedly help to promote the type of open vegetation they prefer. The clearings where gorillas nest, feed and rest are never allowed to regenerate through secondary forest succession to primary forest because of the regular destructive effect of gorilla activity. In some parts of the Virunga Park gorillas' repeated use of vernonia trees is destroying that type of vegetation cover in favour of herbaceous ground cover. The gorillas are obviously unable to keep the whole of their large ranges under such intensive utilization, but they can maintain a system of fairly open trails linking their favourite clearings.

Of course, where human agriculturalists have practised their slash and burn shifting agriculture they leave behind them large patches of regrowing secondary vegetation which is ideal for gorilla occupation. It is in such artificial secondary forest that the Kahuzi-Biega gorillas spent most of their time and the same phenomenon is true throughout the whole range of the gorilla. Spanish zoologist Jorge Sabater Pi[4] observed lowland gorillas in Rio Muni in West Africa and described exactly the same preference for secondary growth within the primary forest. These secondary forest patches are thick with the common wild ginger herb which is the most important food item of the Rio Muni gorillas, who eat its leaves, stems, roots and fruits. The fruits need to pass through the digestive tract of a large animal in order to germinate properly and Sabater Pi concluded that gorillas and wild ginger had had a long and mutual co-evolution. The ginger thickets supply the gorillas with the bulk of their diet whilst by destroying developing tree saplings and spreading ginger seeds the gorillas guarantee that large areas of secondary forest remain covered with this type of vegetation.

There is something slightly strange about the gorillas' preference for open secondary forest and ground layer vegetation. Over ninety per cent of the actual distribution of gorillas is in lowland rainforest, which is a predominantly closed primary forest, in which the bulk of available foods, namely the great wealth and variety of leaves, flowers and fruits, are up in the tree canopy and in which the herb layer is very sparse. Rainforest fauna are typically arboreal in their habits to exploit these tree-borne foods. If this is really the gorillas' original habitat, why is the gorilla not more arboreal? Why does he make no attempt to exploit the main food sources of his environment? Why does he prefer the scattered patches of new secondary forest which occur along river banks, on shifting soils, where a treefall has made a temporary gash in the canopy or where by repeated effort he has kept the vegetation open? The gorilla's is such a peripheral and limited ecological niche that there must be some good reason why the canopy of the primary forest is out of bounds.

Again perhaps Sabater Pi can supply the answer.[4] Over almost all of the gorilla's range it lives side by side with the more agile chimpanzee. The chimpanzee is less exclusively a rainforest animal than the gorilla and occurs in quite arid areas in some parts of its distribution. But where it does occur in rainforest it is very much a canopy feeder. Sabater Pi found that the spatial exploitations by gorilla and chimpanzee were highly complementary, the larger apes feeding mainly on the ground while the chimps fed in the trees, the gorillas occupying the secondary forest while the chimps made most use of the taller, primary forest. The gorilla's way of life makes far more sense if we see him as having evolved against a background of chimpanzee competition, where the smaller ape has proved to be ecologically more effective at utilizing the canopy forest layers and the gorilla has survived only by specializing as a secondary forest browser. Perhaps the chimpanzee is not so clown-like as he appears.

How does our survey of cousin gorilla relate to our findings on the lesser apes, the gibbon and siamang? Those stable, small families, so different from the monkeys but rather like human families, are not repeated in gorillas. Gorillas, by contrast, live in large superficially monkey-like groups which are less stable in their composition than gibbon families, do not contain monogamous partnerships and are not so strictly territorial. There are similarities in the type of strong bonds that link one gibbon partner to another, and those by which the central silverback male gorilla is linked to each of his females, but there are few clear characteristics that are common to each of these apes and also to man. The apeliness that links us all is proving elusive and we must move on to examine the remaining apes to see what greater clarity or confusion they bring to our quest.

CHAPTER 4

Chimpanzees

TANZANIA 1965-66

27 April 1966

My thatched house clung at a crazy angle to the steep slope that dropped away to the gravel beach below. The first light of day danced on the surface of the calm lake, framed in the silhouette of soft-leaved acacias. I closed the door and followed the narrow track through the dew-wet, chest-high grass. Drowsy baboons coughed and scuttled away while the crisp fronds of the wild oil palms rattled overhead in the gentle breeze. I turned on to the main path from the lake shore and headed up the Kakombe valley towards the chimp camp. Shrouded in the darkness of the tall gallery forest lay the broken poles of the old camp huts. It was strange to think that only a year earlier people had lived here but now it was a ghostly place, haunted by bushbabies and porcupines. The path circled the broad, buttressed iron-wood tree where the chimpanzees liked to drum loudly in their displays and I caught the faint smell of ripening bananas from the metal store house. The tall silver-barked dudu tree stood stark as a sentinel against the brightening sky, and the dark shape of the hill we called 'Sleeping Buffalo' loomed up on the far side of the valley where thickly forested gulleys fell steeply to join the gallery forest of the Kakombe stream.

Rounding the corner I reached the two metal houses which were the new Gombe Stream Reserve chimp camp. Caroline and Sally were up, ready for the first chimpanzees to visit the camp. The banana boxes were loaded last night, ten bananas in each. Every chimpanzee that came to camp was allowed to remove the contents of one box and this provision encouraged the animals to visit the post frequently, so that their behaviour and interactions could be observed and recorded. The girls were kept virtual prisoners in the camp by the regular visits of the sixty chimps that had learned about the ever-fruiting banana 'grove' in the Kakombe valley.

The first black shapes of early chimps came up the slope, but my gaze was fixed on a clump of tall miombo trees where the young, thirteen-year-old male, Pepe, had nested the previous evening. I intended to follow Pepe today.

It was still early days in the Gombe station built up by Jane

Goodall and Hugo van Lawick for research on wild chimpanzees. Most of the observation at this time was confined to chimps visiting the feeding station. A few brief observations had been made on animals encountered in the forest but most of the chimps were shy about being followed. In my guise as an entomologist I was not tied to camp like the girls, and during my wanderings I had often met chimps. Pepe was the least shy of all. I had already followed him for several hours and, far from trying to lose me as other chimps did, Pepe would sometimes glance over his shoulder and wait for me to catch up as though he actually enjoyed my company. I knew that with a bit of persistence I should be able to follow Pepe for a complete day. I wanted to see how far a chimp travelled, his daily routine, who he met in the forest and what he ate.

Attracted by the grunts and squeals of the early arrivals enjoying their bananas, Pepe came loping down the slope towards the camp, gathering speed as he travelled.

5.45 Pepe arrives in camp with a tremendous display, hair erect to enlarge his appearance, standing upright, stamping and beating his chest. He empties a box of ten bananas and carries his booty into a wild custard apple tree to eat.

Dramatic arrival displays are the normal way for an adult male to impress a group with his presence, but chest-beating was one of Pepe's specialities. Although a universal element in the displays of male gorillas, chest-beating is rarely seen in chimps; only one other chimp in the whole Gombe population gave such performances and he only with one hand. Pepe's magnificent displays were an important factor in his rising status in the male hierarchy.

5.55 Pepe has nearly finished his bananas when two old males, Leakey and Huxley, start a brief but noisy squabble. Upset by the dispute Pepe gives another display, brachiating through the trees, chest-beating, stamping across the metal roof of the observation hut and then running across the clearing, dragging a large branch behind him. Having demonstrated his displeasure, he returns to his tree and resumes feeding on the banana skins he is still holding. He chews and sucks on the fibrous wodge of skins.

A group of chimps out of sight on the far side of the valley give a loud series of modulated contact calls and Pepe, his mouth still full, answers with a high-pitched call of his own.

6.06 Pepe leaves the camp with the large, though still low-ranking male, Humphrey. They file along the path up the valley where they pass top-ranking male, Mike, and his friend, J.B., who are heading towards camp. Pepe climbs into a small tree, makes a nest and lies in it still chewing his wodge of banana skins. Humphrey goes a little further before he too nests. Both chimps rest for over an hour till the sun comes up over the mountains at the top of the valley.

Pepe rouses and scratches and when Humphrey parades beneath his tree he climbs down to join him and they return to camp. Pepe enters camp followed by the small, low-ranking male, Worzel, whom he has met on the path. Pepe climbs a tree and shakes a branch as a summons to the childless adult female, Gigi, who is in oestrus with swollen pink bottom. Gigi responds to his command and squatting on a branch allows Pepe to mate with her.

Pepe then goes to greet his adolescent eight-year-old sister, Miff, and they groom each other, only to be joined by Gigi, the adolescent female Pooch and the sad orphan, Merlin. Merlin is three years old, the younger brother of Pepe and Miff. Their mother Marina died a year earlier and Merlin had since become a thin, rather sad and neurotic chimp who no longer played with the other infants but tagged along with his sister Miff, who tried to be a second mother to him, sharing her nest with him each night.

It is the busiest hour at the chimp camp; single animals, family parties, twos and threes, drop in to see if there are any bananas in the offing or simply to mingle with the other chimps in what has quickly become the social centre for the ape community. Several adult males make dramatic charging displays, terrorizing their juniors, chasing and mating with Gigi and mother Melissa, who are both pink at present. A cluster of mothers groom together on the hillside above the stream and their babies get together for rough-and-tumble games, only to dash for the safety of the trees when the next party of big males arrives and dashes through the camp, hurling rocks and branches and shrieking the place down.

After an hour of camp excitement, Pepe has had enough and sets off up the valley, followed by sister Miff and little brother Merlin. They pass mother Passion and her infant Pom, who are heading towards camp. Pepe finds a small tree of ripe palm nuts, collects a mouthful and carries an armful of extra nuts to another tree further up the hill where he can eat them in peace.

He has about twenty nuts altogether. Miff goes to the palm tree and picks some nuts for herself. Merlin picks about three, then goes up to join Pepe. He tries to beg more from Pepe instead of picking them himself, but is out of luck. Pepe chews the nuts round in his mouth, scraping off the rich, oily flesh. Every now and then he takes one out for examination. If it is finished he drops it, if not he holds it in one hand. As he discards nuts from his mouth he replaces them with better ones.

8.40 Miff has finished her nuts and goes up into a custard apple tree. Merlin is making soft, whining noises and is still begging with outstretched hand. This time Pepe lets him take a nut from his mouth.

8.45 Pepe moves to another tree, spitting out a nut as he travels. Miff is lying in a small, leafy day-nest in a tree. Pepe sets off up the hill; he goes very fast and is soon out of sight; Miff and Merlin follow way behind.

8.50–9.00 I have found Pepe again in a small tree on top of Peak Ridge eating milk pods. These are very hard woody shells, two to three inches long, enclosing four soft white beans in sticky, white latex-covered envelopes. Pepe bites the pods in his mouth, spits out the hard shells with most of the sticky folders, eating only the white beans. I can hear munching in another bush behind us, so assume that Miff and Merlin have caught up now and are also eating milk pods.

Peak Ridge was a narrow rocky ridge covered in grasses, small trees and bushes. From there I could survey the entire Kakombe valley, from the mountain we called Molo at its head right down to where the stream filtered through a thick stand of bamboos into the bright blue of Lake Tanganyika. It was a fascinating valley of open grassy slopes, patches of mixed woodland, stands of tall miombo trees and steep gulleys of lush gallery forest. Wild oil palms had become well established on the lower slopes, and way below me a palm nut vulture glided in from the shore on broad black and white wings to settle on its nest.

It was beautifully peaceful up on the ridge, and I found it almost impossible to imagine the dozens of tense ape dramas that were being enacted beneath the inscrutable leafy canopy below. The valleys on either side still looked lush and green but there would be little rain now for several months. The grass would soon begin to brown off and by the end of the summer fires would sweep across many of the open ridges of the reserve before the November rains could bring new growth again.

At midday Pepe is still eating milk pods. Miff clambers into another tree and starts eating an old brown gourd. Half-way through it she notices a better green gourd and drops the remains of the first. She picks the hard green fruit and carries it to a tree where she bashes it about fifteen times on the trunk until it cracks. She bites it open and starts to eat the contents, spitting out unwanted pieces of shell and seeds.

12.15 Merlin is sitting almost asleep in a thick shrub, his head resting on his leg. Miff and Pepe are still munching through the milk pods. I have been counting the fallen shells and Pepe has had well over five hundred by now.

At last Pepe descends and starts off down the hill to the north – into the next valley. He passes a small protea bush and picks a large fluffy flower and a fat green bud to eat. This is quite exciting, for it is the first time this item has been documented as a chimp food. Pepe continues into the Kasakela valley, climbs into a large tree and lies down in the crotch of a branch about twenty feet off the ground.

The family enjoy a leisurely hour of resting and social grooming, working carefully through each other's fur with their index fingers and plucking off pieces of dirt or flakes of skin with their lips.

But renewed hunger eventually ends their rest and the three chimps set out for the milk pods yet again. Even little Merlin manages to crack open a few of the hard pods for himself.

14.45 Pepe comes down from his tree and the whole family moves off down the slope deeper into the Kasakela valley. A large male baboon appears and makes as though to dash at Merlin, but Miff rushes up and chases the baboon away. I can see the adult male chimp, Huxley, resting in a tree nest about two hundred yards away along the side of the valley. He appears to be alone. The chimps seem to have noticed Huxley also, and Pepe is leading the way straight towards him.

15.05 All three chimps climb up into a tree near Huxley's and start to feed on spikes of small yellow flowers. They don't eat many although there are plenty of flowers there, but seem to take just a few for variety. Miff and Merlin climb down and move across to Huxley's tree where the female Pooch has also appeared. It is another of the yellow flower trees but again there is very little feeding activity.

15.20 Pepe and the others climb down from their trees and set off in single file along an old animal trail across the slope of the hill with Huxley in the lead. Pepe lags behind.

15.30 The animals have moved further down into the valley and are feeding among great tangles of secondary vegetation on a wealth of green leaves and vine shoots. Creepers have formed a dense carpet and I am walking on a springy mattress of greenery, several feet above the ground. Pepe is still separate from the other animals but within sight of them; he is at the top of a small bush eating leaves. Suddenly he heads off into very dense undergrowth where I can only follow on hands and knees. He finds a small termite mound and bites it open with his teeth. Wriggling masses of white termites are exposed and he licks these up with apparent relish. Quietly he slips away into the dark thicket.

15.55 I have found Pepe again. He is motionless, staring straight ahead. There are slight stirrings in the undergrowth where some animal is breaking off leaves. Carefully and quietly Pepe moves towards the noise but it turns out to be just another chimp, an old male, Worzel. Pepe stands up, hair on end and does as violent a display as his confined space permits. He shakes vines, rocking them to and fro, then dashes down the hill swinging through the trees and vines, beating his chest. The other chimps are moving away back up the hill. It starts to rain and the whole group, including Worzel and Pepe, rush off too fast for me to keep up in such thick underbrush.

I follow in the direction they have disappeared but it is difficult to hear anything in the rain and there is no sign of the chimps.

16.15 I sight the group in a fig tree munching away at leaves. We are deep down in the Kasakela valley now, quite near the stream bed.

16.20 A female chimpanzee, just coming into oestrus judging from the size of her sexual swelling, climbs up to join them. It is the female Pallas who is only an infrequent visitor to the chimp camp and still very shy of human observers. The whole group now consists of seven chimpanzees – males Huxley, Worzel and Pepe; females Pallas, Pooch and Miff, with young Merlin.

We hear chimps calling up near the Peak ridge we had crossed earlier in the day. My group reply immediately with their own chorus of barks and pant hoots, then resume eating. It is pouring with rain. Pepe rubs his back on a branch and plucks a bunch of wet leaves to wipe his face. It is still covered in the sticky latex from his feast of milk pods and he seems to be trying to clean this off with the leaves. Shy female Pallas sees me and hurriedly climbs down from the tree, intending to run away silently on the ground. This is typical behaviour for wild chimps seeing a human. Unfortunately for Pallas, the other chimps are not frightened of me and continue to feast on fig leaves. Pallas is caught in a dilemma. She wants to run away but she does not want to leave the other animals. She compromises by climbing into a small bush and watches me suspiciously, uttering rather pathetic little hoots.

The other chimps continue to feed for an hour before Pepe finally leads the way down the tree and the whole group sets off in single file up the slope to the Peak ridge.

17.30 They are in a clump of tall trees just under the ridge between the two valleys. I can hear the snapping and bending of branches as Pepe starts to make his nest for the night. It is time for me to retire too and I hasten back over the ridge and down into the Kakombe valley, hoping to reach camp before it gets dark.

Pepe's day tells us much about the nature of chimpanzee society, a society which is far more complex than that of the gibbon family or gorilla group. Pepe's route and activity pattern for the day was unique and individualistic. Sometimes he was alone; sometimes he travelled with his siblings in a family unit; sometimes he enjoyed exclusively male company, not the dominant males, but other low-ranking males such as Humphrey, Worzel and Huxley, more in keeping with his own youthful status. He had also mingled with a group of mothers and families, and finally joined and nested with a mixed group of males and young childless females. He had mated with one female, though not exclusively, had groomed with four different animals, displayed at more and avoided yet others. In all he had interacted with twenty-seven different chimpanzees in the course of one day. Next day he would probably meet as many chimps again, many of them new contacts. Certainly Pepe spent more time with his family than with anyone else, but he moved easily through the various circles of the chimpanzee community in a manner quite distinct from anything we have seen in the other apes or in monkeys, but very similar to the social complexities common in human communities.

The courtship and sexual behaviour of chimps, however, seem very foreign to our own species. Female chimpanzees indicate their sexual condition by strong visual and olfactory signals. At the height of monthly oestrus the female's bottom becomes grossly swollen and flushed pink. Adult males were only interested sexually in the females while they were pink, but during this time they attracted great attention. Males of high status summoned such pink females with a brief threatening courtship display, by staring straight at the female and shaking a branch or vine. This message was quite clear and if she did not approach the male, presenting her bottom for his inspection and probable copulation, she was liable to be attacked. Copulation was a brief affair with the male mounting from behind, and taking only three or four seconds to reach his climax. If young animals were present, especially the young of the female concerned, they would rush up to interfere with the copulating couple, barging between them and forcing them apart. Males showed extraordinary tolerance at such interference but the behaviour remains difficult to interpret; the youngsters showed every sign of being jealous of the attention the male was giving to the female, but the evolutionary significance of the behaviour is probably that by disrupting their mother's sexual behaviour they may sometimes prevent a pregnancy that would eventually rob them of their mother's attention.

Male chimpanzees have a tremendous sexual capacity. They have very large testicles for their size and an enormous output of semen. An adult male may copulate as many as twenty times in a single day, and since there are always some pink females in the community, the males enjoy sex regularly.

From the human viewpoint, chimpanzee sex has become enormously cheapened. It is very brief and has little meaning or importance to the performers. It is often no more than a polite way of greeting another chimp. A male who is feeding will continue to feed throughout the performance and several males may mate, one after the other, with the same female.

Curiously enough, despite all this volume of sex, the chimpanzees who regularly came into camp as pink females, and were mated by all and sundry, rarely became pregnant. The sex act seemed to serve more of a social than a reproductive function. Female chimpanzees appeared much more likely to become pregnant if they avoided the main community and went off on a 'honeymoon' consortship with a single male.

About two months after my arrival at Gombe, when I found myself for a few days the only person manning the post, the handsome male chimpanzee Faben arrived in camp. The same day a female called Mandy arrived. Both Faben and Mandy had been absent from camp records for a few months, and when I checked back I found they had both disappeared at about the same time. I suspected they had been away together and when Mandy was subsequently noticed to be pregnant I had little doubt that Faben was the father.

A few months later we noticed that the female Olly had not been seen in camp for several weeks, although she was normally a very regular visitor. Attendance charts showed that David Greybeard had also been absent over the same period. Perhaps they too had gone off together and I kept an eye out for them on my entomological trips away from the home valley.

A month later Olly and David had still not been seen and we were beginning to get rather worried about them. Then, on 11 April, several miles to the north in the Mtumba valley, I recognized a small chimpanzee who was foraging in a pocket of forest near the lake shore. It was Gilka, Olly's four-year-old daughter. I knew that Olly could not be far away. Sure enough, I found her and David Greybeard feeding among a thick tangle of vines. It was with considerable elation that I reported to the camp that the prodigals were in good health. A month later Jane Goodall and her husband, Hugo van Lawick, returned to Gombe from a filming safari. David Greybeard was Jane's closest chimp friend. Jane and Hugo set off towards Mtumba to search for him. They went there in vain, but I am sure that David must have seen them and remembered the happy banana-rich life of the chimp camp for the next day he returned with Olly and Gilka. Olly was pregnant.

During the ten years since I left Gombe Stream, the importance of these consortships has been repeatedly confirmed.[1] Males lure single females away with them. The pairs avoid the activity centres of the main community, indeed several other consorting couples have used the very same quiet little valley where I found Olly and David. After an absence of a few weeks or even months the pair return to the main group, when the female is usually pregnant.

The curious thing about this honeymooning is that there is little correlation between a male's status and whether he fathers infants. In monkey groups there is competition between males for status or dominance, and the pay-off for being top of the hierarchy is more easy access to the females so that it is the dominant males who father most of the infants. Among the Gombe chimpanzees there was certainly competition and status rivalry between the different males, but the advantages of status were not so clear. A dominant animal could take food from a subordinate. In camp where animals sometimes obtained armfuls of bananas this would be an advantage, but in the wild animals picked their food one bit at a time. It is as easy for a dominant animal to collect his own food as to lift morsels off his subordinates.

Dominant animals do usually get first option of pink females but they do not maintain exclusive rights to these prizes. Many different males accompanied females away on consortships and several of these males were quite young, still low-ranking animals. Although status may be a slight advantage in procuring females, reproductive opportunity is much more equal for male members of chimp communities than in any other group-living primates. Neither the male Mike who was dominant animal throughout my stay at Gombe, nor his predecessor, Goliath, had shown much interest in the opposite sex.

Chimpanzee social organization approaches that of a commune. The males in each commune are linked by their mutual need to defend their community range. To ensure that the males do co-operate in this task, it must be evolutionarily worthwhile for each individual to help the whole community. Since each male chimpanzee has equal reproductive opportunity his chances of breeding are far better if he stays within his parental community than if he tries to make it on his own. Since no male can distinguish his own offspring among the infants of the group, he must co-operate fully for the benefit of the whole community, to ensure that his own offspring get the best possible chance of succeeding in their turn.

Sexual jealousy does exist between different males within the community but they cannot afford to tear each other to pieces in their competition. Consequently, chimps have very stereotyped displays and controlled attack behaviour which enables them to settle disputes with a minimum of injury. The fairly stable dominance relationships which are clarified by the elaborate system of greetings, and gestures of ingratiation and reassurance, effectively reduce fighting to a minimum. The dominant male does not lord it over the group like an alpha male monkey. He cannot be recognized by appearance as is the case in most monkey troops; he is not the biggest animal nor the one that has the longest hair. He does not monopolize the females. He does, however, act as leader in fighting off other chimps, baboons or any other threats, and in return gets some comfort advantages in frequency of being groomed, relative absence from the permanent chimp fear that someone bigger is about to arrive and attack him, and some advantages in feeding situations. Moreover, as chimps enjoy a long life of about thirty years but remain dominant male for only a few years, most of the high-status males within a community have either already enjoyed the top spot for a while or are likely to do so at some time in the future.

The whole Gombe chimpanzee population was composed of more than one community. Some forty-five chimps visited the camp regularly while another fifteen were recorded occasionally. These irregular visitors usually entered the Kakombe valley from the south and we came to think of them as the southern group. One large male, Rix, appeared to be the dominant animal of the southern group and was treated with great respect by the home group males. Even Mike would not challenge the newcomer.

Rix got into an argument with three of the home males. Their numbers gave them confidence and they chased after him. Rix's bluff had been called, he fled, and thus encouraged the trio hurtled behind, jumping on and thumping him, pulling his hair and generally knocking him about. Screaming, Rix made his getaway with hair flattened and face grimacing. He was never able to pull rank again and on future visits to the camp he had to play subordinate to Mike and the other big males.

The artificial supply of bananas at the camp was enlarging the home community by luring animals in from the surrounding groups. McD was a young sub-adult male who regularly visited the chimp camp, yet did not show any particularly close relationships with any of the other animals. Then towards the end of 1965 we had a new visitor, a juvenile female, Lita. Lita showed an amazing facial similarity to McD, but we could not be sure of any relationship till one day, several months later, a new female with a baby arrived in camp with a party of other southerners. The resemblance was unmistakable. Jessica was almost certainly McD's mother, and sure enough McD and Lita both joined her and all had a family grooming session. McD's origins were suddenly so much clearer. In the two years after I left Gombe Rix died accidentally in a tree fall, and McD and several other animals succumbed to successive epidemics of polio and flu. The survivors of the southern group and camp group merged together. It was only several years later, when banana feeding was drastically reduced, that a social bifurcation took place and led again to the existence of two clear groupings.

To the north, the Mtumba valley supported another small chimp community and I sometimes met members of this group on my travels. They were wild, shy creatures who fled at the sight of me, crashing out of the trees and hurrying away silently along the ground. In the southern part of the reserve there was yet another large chimp community.

We now know that male chimpanzees usually remain in the community of their birth but that young females frequently change from one group to another, thus maintaining outbreeding within the population. Friction between communities may be fierce, with screaming battles when members of a neighbouring community are found trespassing. The males share responsibility for guarding the community range and sometimes a party of three or four of the home males would return to camp sporting fresh wounds. Usually females with babies avoid such battles, for the fighting is fierce and there may be serious accidents. Shortly after my arrival at Gombe Stream Sophie entered camp with her brand-new baby. The other chimps gathered around, admiring the new addition and shortly afterwards Sophie headed up the valley. Later that afternoon we heard terrible screams and shrieks of chimps fighting. Three males returned with fresh wounds but they were not the only casualties for next day Sophie, too, returned, but she had lost her baby. We could never be certain what had happened, but subsequent observations[2] have confirmed that male chimpanzees sometimes attack strange females, killing and even eating their babies.

The killing and eating of other animals was fairly commonplace among the Gombe chimpanzees.[3,4] Most primates take small quantities of invertebrate protein which may be an important part of their diet. In feeding on ants, termites, wasp grubs and caterpillars, chimpanzees do nothing untoward, but their tremendous speed and strength enables them to tackle much larger game. Prey of the Gombe chimps included the young of bush pigs and bush bucks, monkeys and young baboons. I was fortunate to witness a hunting party at work. A strange whistling sound a short way up the valley brought all the chimps hurrying from camp and I rushed after them. A chimp had cornered a red colobus monkey in an isolated clump of trees, and the other chimps were spreading out in a spooky and serious manner to cover all possible exits. But this monkey was lucky, spotted a weakness in the hunting ring and flung itself across a wide gap to escape. Other animals were not so fortunate. Chimp Rudolf seized a young baboon close to camp and bashed it to death on the ground. The chimp carried his prize into a clump of tall miombo trees to eat. We watched, fascinated, as excited chimpanzees crowded round Rudolf, whose status was temporarily raised as a result of his bold deed. Even Mike and J.B. showed him due respect and did not dare to steal the corpse. Rudolf began to rip the carcass apart, tearing strips of rich red meat. It was only hours later, when he had already gorged himself, that Rudolf permitted the other chimps to share the battered remains.

Such hunting of baboons was in marked contrast to the normal relationships that existed between the two species. Baboons and chimpanzees frequently met at camp and interacted with relative peace and tolerance. Young baboons and chimps played happily together and although there was competitive friction between adults over food, it was usually the long-toothed baboons who dominated the chimps with their aggressive behaviour. The relative tolerance and even cowardice by chimpanzees completely belied the fact that these same domineering baboons formed their chief animal prey species.

Female chimpanzees are not usually strong enough to rip open a monkey carcass, so the hunting of large prey is a predominantly male activity. Conversely it is the females who show the greatest patience in catching insects and it is in this field that we see the highest degrees of chimpanzee skill and inventiveness in the use and manufacture of tools.

In parts of West Africa chimpanzees use digging sticks to break into the earth mounds of termites so that they can feast on the exposed inmates.[5,6] Perhaps the drier weather and different earth texture makes Gombe termite mounds too tough to open, but the chimps have developed another way of obtaining the juicy contents. They go fishing.[7] During the rainy season the sexual forms of the termite colonies emerge in fluttering swarms. Chimps and birds often crowd round to catch these tasty morsels as they appear, but chimps can also locate where the emergences are being made, often days before the winged insects are ready to fly. The chimps excavate these exit tunnels with their fingernails. The intelligent chimp has learned that if he pokes a long grass stem into the tunnel the soldier termites will attack and bite on to it. Then all the chimp has to do is withdraw the stem and pluck off the harvest of delicious termites.

During a rainy spell in 1966 I watched aged mother Flo and her infant son Flint working on the large termite mound below the main camp. Flo was an old hand at termiting, she knew exactly where to work and when to change holes when she wasn't catching enough. Flint was watching closely; he took hold of one of Flo's discarded fishing stems and tried for some termites of his own, but he was still very inept, holding the stem much too short. It would take a lot of practice, maybe two more seasons, before he would actually succeed in capturing termites for himself; till then termite fishing was just a game to watch and copy.

It was easy to see how such a useful discovery would be passed on and perpetuated in the culture of the community, but it must have been a genius of a chimpanzee who first perfected the technique of termite fishing.

The Gombe chimpanzees showed several other examples of tool use. They caught biting safari ants by dipping a stick into a marching column of these ferocious little creatures, then licked them off lollipop fashion. They attempted to prise open banana boxes with larger sticks or probed the boxes with fine twigs, for by sniffing the ends of these tools they could tell if there were bananas inside or how ripe they were. Infant chimps used sticks and other objects as toys, to wave about, throw and manipulate in a hundred different games. Chimps used bunches of leaves to wipe themselves if they were soiled with faeces or blood and we have already seen how Pepe used wet leaves as a flannel to wipe the milk pod latex from his face.

A few of the chimpanzees also used leaves as a drinking sponge. Chimps sometimes drink water from natural water traps in tree-holes, scooping the water with their hands like gibbons and siamangs. But some of the Gombe chimps found that it is quicker and more efficient to use a bunch of chewed leaves as an absorbent sponge, dip it in the water-hole and squeeze its liquid into the mouth.

Chimps also show a high degree of skill in aimed throwing. In their normal arrival displays adult males frequently toss loose vegetation, sticks or even rocks aimlessly as they rush past. Some chimps carried stones for several yards in their displays before throwing them at a specific target; others threw missiles in an even more deliberate and calculated manner. I once saw Humphrey pick up a large stone and amble casually towards a baboon, turning his hand so that his intended target could not see the missile. The baboon was quite unaware of the impending attack. Humphrey ambled past without the slightest sign of interest, then just as he drew level he swung the stone arm viciously at the baboon, scoring a bull's-eye, and the frightened victim rushed away with a yelp, not sure what had hit him. Old Flo's adolescent daughter, Fifi, was another malicious stone-thrower. She was more prone to throw a handful of pebbles than a large rock and had a particular hatred for the adolescent male Hornby. Several times when I watched Fifi resting near camp I saw her suddenly start gathering a little pile of stones and I would know that Hornby must have arrived.

Throwing missiles is not just a way for bad-tempered chimps to settle old scores, though. It probably serves a much more vital function, a means of protection from predators such as the lion and leopard with which the chimp shares much of its distribution. The Dutch primatologist, Adriaan Kortlandt, conducted an experiment[8] in West Africa in which he exposed a stuffed and animated leopard to a party of wild chimpanzees, and filmed the chimps' initial alarm then angry attack of the model. The apes picked up large sticks and made repeated individual charges at the model, throwing the sticks as projectiles and actually clubbing the leopard with them. I am quite sure that the chimps were not fooled into believing that this was a real leopard. Nevertheless, the experiment demonstrated that chimpanzees will use weapons to attack potential predators and helps us understand how small bands of chimps can travel with relative lack of caution through leopard-infested forest. It would obviously have to be a pretty hungry cat who would dare attack these black screaming imps, who not only have dagger-like sharp, fierce teeth, tremendous strength and speed, but hurl rocks and use clubs to boot.

The importance of culture in chimpanzee communities was only too evident at Gombe. Chimpanzees copied and learned from other animals. Each chimp could benefit not only from his own experiences and those of everyone else in the community, but also from the lessons of preceding generations whose discoveries had been culturally propagated. Moreover, the Gombe chimps showed innovations in their behaviour. One year it became the fashion to nest in palm trees;[9] another year it was more fashionable to hunt baboons than the more traditional red colobus monkeys. Animals mimicked what they saw before they understood its function, but they were usually quick to grasp the significance of their performances. They imitated not only the activities of other chimps, but any human action that caught their attention. On one occasion I spent some minutes watching a jumping-spider stalking and immobilizing a robber-fly on a tree trunk. Fifi had noted my intense interest and as soon as I moved away she dashed across to the trunk and scrutinized the spider as closely as I had done. Not finding anything to merit such attention she eventually ambled away, but not before throwing a puzzled look in my direction.

Fifi was a genius among chimps, an innovator, always making new discoveries. She was a real investigator, inspecting and fiddling with everything. If she met a dead animal on the trail she just had to pick it up, open its legs, sniff its rear end and poke it about. If she found a bent piece of wood, she would play with it and examine it thoroughly, as though trying to discover a use for its particular physical properties.

Fifi was one of the experts at opening banana boxes, unscrewing the bolts till the last thread, waiting for the right moment when no one was looking and there were no males about before giving the final twist, opening the box and hurrying up a tree with an armload of bananas. Fifi was also the inventor of the nest roof. One day I watched her make a day-nest in the crown of an oil palm tree behind the camp. Long after it was completed by normal standards, she continued adding to and amending her nest. She bent palm branches over her head, tucking their ends into the simple weave. She added more and more of these overhead branches till she had a cave-like structure with a roof and three sides. She was obviously rather pleased with her construction, for she returned to it again the following day and continued with minor improvement schemes. Normally chimps use a new nest each day. Her novel nest design was merely the result of a game, but under certain weather conditions it could afford much greater shelter than a normal nest and, should the benefits ever exceed the extra labour involved, Fifi would know how to cope and other chimps could learn from her.

The advantages of culturally-acquired behaviour are enormous. New patterns of behaviour can be acquired fast and dropped quickly again if they become inappropriate. It would take hundreds of generations of consistent evolutionary advantages for such behaviour patterns to be acquired genetically and they could never be so adaptable as in the opportunistic chimpanzee. Nevertheless, there are dangers inherent in a dependence on learned behaviour. Faults of memory may lead to both errors and delays in performance, compared with animals in which the appropriate response is instinctive. Also there is great hazard during the learning period. An instinctive response is given correctly the first time it is required, but the young chimpanzee runs a real risk of only discovering that a poisonous berry is not good to eat or that snakes are not to be sat on after he has eaten the berry or been bitten by the snake.

The chimp's cultural repertoire depends upon the inventive, inquisitive and exploratory nature of the young animals, but to ensure that curiosity does not kill the chimp the adults develop and exercise a high degree of conservatism. When infant Flint was playing with a large and sluggish cricket, his mother, Flo, pulled him away from his toy and flicked the offending object out of reach. There is little to fear from a cricket; at worst it might exude an evil-smelling oily fluid and certainly it would not be good to eat. No, the real danger lies in the fact that if Flo is not strict about crickets, Flint might one day start playing with a scorpion when her back is turned. Flo also exercised a similar authority over strange foods. In our camp we had oranges, pawpaws, passion fruits, tomatoes and many other foods that any chimpanzee might, and in zoos does, enjoy; but the Gombe animals showed no interest in experimenting with them. Flint might find a piece of pawpaw skin by the rubbish dump, but before he could smell it or taste it, Flo would confiscate it and throw it away. Traditional ways are safest, if not always best, and the rate at which new discoveries can be made by inquisitive youth and spread culturally through the community is severely checked.

The consequences of adult conservatism in chimps can sometimes be mistaken for stupidity, just as youngsters in our own community see their conservative elders as unenlightened. A classic example of this was the chimpanzees' attitude towards the camp veranda. Chimpanzees loathe water and are never so miserable as when heavy rain falls. They assume a standard rain posture, squatting huddled up, hands together in front of their chest and head lowered. In this position all the hairs of the body hang downwards and most of the rain slides off. Chimpanzees will shelter against tree trunks or under foliage, for they have learned that this affords them some protection, but they had never experienced a totally waterproof veranda roof before. Several times I saw chimps who were sitting underneath the veranda roof when a rainstorm broke; these animals assumed the typical rain posture and sat huddled tight. All around they could hear rain beating on the roof and foliage and see blankets of water splashing on to the ground but they were not getting wet. Worriedly they gazed about, then finally unable to accept the novelty of not getting wet when it rained, and unwilling to ponder on the physics of waterproof shelters, they walked unhappily out into the downpour to get traditionally and uncomfortably soaked. It took several months before chimps learned to accept the shelter of the veranda roof, but before they reached the point of actually heading for the veranda when it rained we had to take the roof down so as not to be crowded out.

The chimpanzee is a leisurely animal. At most times of the year he can satisfy his daily needs with only a few hours' travel and feeding, the rest of the day is free for pleasure, social intercourse or plain laziness. Leisure time occurs, as in most tropical animals, in the middle of the day when the temperature is rising and activity is more fatiguing. Mothers and their young form social parties with other families and, while the adults rest or groom one another in the chimp equivalent of social conversation, the youngsters play and explore. To describe chimp games would take a whole book, for they are amazingly inventive. Any clump of grass, a stone, broken branch or tangle of vines might inspire a new variation of classic games: rough and tumble, chasing, fighting, mock displays with branch-waving and rhythmical thumping, throwing and juggling, locomotor games like pirouetting, somersaulting, swinging on a vine or dangling about, carefully grooming leaves, making play-nests, teasing parents, torturing grasshoppers or simply setting about the systematic destruction of a bush. Sexual differences are evident at an early age, male infants and juveniles indulging in more wrestling, displaying and playful imitations of copulation whilst juvenile females already show a great interest in small babies, grooming them and sometimes, if the babies' mothers aren't looking, actually kidnapping babies and carrying them away.

Undoubtedly play in chimpanzees, as in many other mammals, is really a way of learning skills that are going to be needed in adult life, skills of locomotor ability, object manipulation, fighting techniques, display, nesting and mothercare. But whatever the evolutionary reasons for chimpanzees' play, the immediate pay-off is fun: chimps play because they enjoy it and their pleasure reinforces the behaviour. It is in this context of play as an activity performed for the sheer joy of it that we can see the buds of aesthetic taste in the chimpanzee. Here we have a species in which selection has favoured the development of an ear for time and rhythm, vital in giving and recognizing individual pant hoot series but, as a by-product, enabling the playing chimpanzees to enjoy a primitive form of percussion music. Here is a species with a perceptive eye for pattern, vital in recognizing different types of leaf or minutiae of meaningful expression on another animal's face but also allowing chimpanzees at play to indulge in elementary visual art. I remember watching Fifi toss clouds of dust in to the wind and watch it drift away. Goblin came towards her and tried to copy her; he could not do it well but succeeded in tracing patterns in the dust with his fingers. On another occasion Goblin made scratch marks in mud and once even traced the edge of a leaf's shadow with his finger.

As impressive as the chimpanzees' primitive culture and proto-art was their uncanny facility for communication. Like many other apes and monkeys they had a complex variety of vocal grunts and squeaks, as well as many visual grimaces and subtler gestures of posture or outstretched arms. They hugged, kissed, bobbed up and down, waved, beckoned and performed many other meaningful actions. Such behaviour was not merely the performance of stereotyped signals that reflected their emotional state; chimp communication goes much further than this. These signals were given with intelligence and cunning in the full understanding of how the signals would be interpreted by other chimps, and what the consequences of such interpretation would be. A junior chimp would make a play of submissive gestures towards a higher-ranking animal, not because he felt submissive or inferior, but to avoid immediate show-down or allay suspicion from some plan of action he was trying to conceal. When a male chimp sat down and shook a branch whilst staring at a pink female, he was not merely performing some fixed behaviour pattern. He was sending a deliberate and calculated message that he was confident would be clearly understood: 'Melissa, you come here and let me mate with you or you will find yourself being shaken like this branch,' – a common message all the clearer through repeated use.

Even subtler communication went on at Gombe all the time, particularly in the tiny facial movements. Chimps watched each other very closely and could learn a lot from small changes in the facial expression. They could read us humans too and it was difficult to conceal meaning from such intelligent apes, even when one tried.

I particularly remember an occasion when Olly and Gilka were in camp. Olly was a nervous worrier, an over-protective mother, and rarely let Gilka out of her sight. Yet one day I watched Gilka walk straight up to her mother, stare closely into her face for a moment, then turn away and walk out of sight down the hill. Normally Olly would have been after her like a shot, but this time she sat unconcerned and a couple of minutes later Gilka returned safely. She had been down to the stream for a drink. As a scientist I may find it hard to convince the sceptic, but as one who has lived for a year with a community of wild chimpanzees I have no doubt in my own mind that that glance Gilka gave to her mother was a deliberate message of reassurance and read something like: 'Don't worry, mother, I will be back in a moment.'

We cannot fail to be impressed by the chimps' human-like qualities. Surely we must be on the right track in unravelling our own species' origins when we study this intelligent animal. Here we have a community-living animal with a complex, individualistic, though co-operative, social system. Here we have an animal with a well-developed culture, learning from other animals much of its daily relevant behaviour – a species that uses and fashions crude tools – a species that hunts for meat – a species that sometimes even turns cannibal – an animal that can almost talk. Certainly the chimp seems far more relevant to the human condition than either the family-living lesser apes or the group-living gorillas. However, there are still two more great apes to consider before we start jumping to conclusions about man's chimp-like origins. We must continue our tropical travels.

CHAPTER 5

Bonobos

ZAIRE 1975

The third and least known of the African apes is the bonobo or pygmy chimpanzee, which was only recognized as a distinct species as late as 1929. The bonobo is the only ape found south of the Zaire-Lualaba (formerly Congo) river system and has a scattered distribution throughout the rainforest of the Congo basin. Because of political uncertainties and geographical isolation the bonobo missed out in the great surge of primate field studies during the 1960s and has remained something of a mystery animal. Lack of knowledge, however, has never been a handicap to conjecture and the bonobo has variously been described as 'man's closest living relative', 'the most intelligent of all the apes', 'the best living analogue of man's australopithecine ancestors', and 'close to the common ancestor of both chimpanzee and gorilla'.

Obviously the bonobo was an animal that merited serious appraisal and was far too important to be omitted from any comparison of ape behaviour. So I had no hesitation in seizing the chance to see this elusive ape for myself and in February, 1975, I was on a small plane circling the flat, swampy forests of the Congo basin, prior to a precariously steep landing at Mbandaka airport in the Equateur province in Zaire.

I spent several fruitless days checking out localities where bonobos had been reported, but these shy apes are becoming rare now and have disappeared entirely wherever humanity has made inroads into their jungle ranges. I enjoyed exploring some of the Congo forests and waterways but I became impatient that I could not find even a single bonobo sleeping nest. I decided to try my luck further east. A young European couple, Noel and Alison Badrian, had recently set out on an expedition to study bonobos and were still in the field, somewhere north of Boende.[1] It was several months since anyone had heard from them but their last report had confirmed that they had made contact with the bonobos. I hoped to visit them and flew on to Boende. There I managed to locate Père Paul, the Belgian missionary who was handling their mail, but was somewhat disappointed to learn that the Badrians' camp was still another hundred miles away. It took me three days of begging lifts to get as far as the palm oil plantation at Bokali, but the last twenty-five miles had to be covered on foot. I was beginning to understand why letters from the Badrians were so rare. Two villagers promised to act as guides to the remote camp and we set off at dawn on the long walk to the Lomako river. The guides had brought their wives with them and I soon realized why; it was the women who carried the heaviest packs.

For several hours we followed the well-trodden path through luxuriant tall rainforest which reverberated to the territorial morning calls of the black and white colobus monkeys. Small hornbills cackled through the canopy, flapping after insects. Nearer the ground, tree frogs had laid their frothy egg masses among the leaves of bushes overhanging the rain-puddles. Here in these forests of the Zaire basin I could almost feel I was back in South-East Asia. We trod the same leafy soil through the same shady places and glades dappled by thin shafts of sunlight. Just as in Asia, the great tall trees that dominated the forest were draped in lianas and epiphytes. Yet there was something different; this was not Asia, and it was these subtle differences in forest structure that were vital to understanding two divergent evolutionary paths followed by the Asian and African apes.

The forest here smelled quite different. A heavy scent of garlic, emanating from the abundant bofili trees, mingled with the powerful odour of monkey urine to blot out any subtler scents. In spite of the humid conditions and ample signs of forest pigs and antelopes on which to feast, no leeches looped across the forest floor. Termite mounds rose to a breathtaking twenty feet, a sight quite unlike anything I had known in South-East Asia. Above all, however, it was the trees that fascinated me. The dipterocarp trees so characteristic of the upper canopy of the Asian forests were absent here, their place taken by giant members of the bean family. There were no trees here to match the Asian tualangs and jelutongs for sheer height, but there were actually more trees over a hundred feet and thev formed a more continous upper canopy layer than anything I had seen in the Far East.

The middle layers were more open and I tried to imagine how a gibbon would take to these forests. He would find it only too easy to travel through the high canopy by running along the major branches and would rarely need to settle for his own unique mode of travel, brachiation. The very different structure of the African forest may explain why there has been less evolutionary premium on extended leaping and gliding here. Africa can boast only one family of gliding rodents compared with the tremendous array of gliding mammals, reptiles and amphibians that parachute through the Asian rainforest.

At midday we reached a small cluster of bamboo huts, a seasonal hamlet used by visiting hunters and fishermen. We had still a long way to go and from now on the route lay through swamp forest. There was nowhere to sleep between this village and the Badrians' camp, so we did not linger but plunged on into the wet potopot, as the Africans delighted in calling their terrible swamp.

We clambered along submerged branching stilt roots, trying to avoid the deeper, muddier potholes; I was glad of my guides' sure sense of direction. There were no markers or visible paths and the potopot would have been an awful place to have got lost. It was not a stagnant swamp as I had expected; instead everywhere was bathed in a few inches of fast-flowing clear water, the colour of tea. Progress was slow and we were sometimes up to our thighs in water. It was certainly not my idea of ape country.

As we approached the Lomako river I noticed large numbers of elephant bones scattered everywhere. My guides told me that this was an area often used by ivory poachers but that most had been caught and arrested now. I hoped so, as the numerous remains of these giant animals painted a sad picture of most callous slaughter.

Eventually the swamp merged with the broad waters of the Lomako river. The guides found a small boat to take us across the deeper water, but it was a crude and wobbly craft and we would have been just as stable and got no wetter had we straddled a floating log. The water level was so high that I could not even see where the river banks should be and we crossed what looked like an enormous and eerie lake scattered with small islets. Ducks and egrets watched us warily as we paddled past. When we reached shallow water again we abandoned the boat and marched on through slightly drier forest towards the Badrians' camp.

I had thought that some of my own research camps had been in remote spots but this was ridiculous! There seemed no end to the terrible journey. We had been on the go almost without stopping for ten hours. What sort of dedication or madness could induce anyone to put such an expanse of quagmire between themselves and the rest of the world? I was as keen to meet the Badrians as I was to see bonobos!

We splashed along for another half-mile then, suddenly, we were there; a pair of palm thatch houses beside a stream and two surprised Europeans, not at all accustomed to seeing visitors.

I introduced myself and handed over the mail I had collected for them in Boende, including my own request to visit them! Alison was Irish and Noel was born in South Africa but both had studied in England. Rather than find jobs and settle down, they had decided to first make a break into the subject that really interested them, wildlife. With no previous experience in this field and no proper sponsorship, they had found it impossible to raise funds for their project but had been so determined that they had used their own savings to embark on their bonobo expedition. I was greatly impressed by their courage and determination.

Two months earlier the bonobos had been regular visitors to an area not far from the camp and Noel and Alison had been able to watch them frequently, but now they seemed to have moved further away. Noel was recovering from a serious bout of hepatitis. He had been too weak to go into the forest or to cross the swamp in search of medical help. Alison had bravely carried on their study and had been able to keep in contact with a group of bonobos but these animals were now about four miles away from camp. With their own African guides to help them, Noel and Alison were able to work separately and thus double their chances of finding and watching the elusive bonobos. Noel now felt much better and we decided that I would accompany him on his patrol the following morning, while Alison checked out another part of the study area.

It poured with rain in the night and the forest was still damp and dripping when we set off at first light. Benjamin, the guide, hurried ahead, stopping occasionally to listen then gliding silently onwards. Everywhere we found armies of fierce-jawed safari ants which ran up our legs to bite at our flesh and clothing. There were so many things that I wanted to examine, so many things that were new to me. Scarlet, cactus-like flowers sprang leafless from the forest floor, parasites on the roots of other plants. We passed several curious clearings where shiny black ants had killed off all the vegetation around their host trees, whose swollen leaf stems offered refuge, breeding chambers and even sappy food. Most exciting of all, however, were the abundant signs of bonobos.

The whole forest was a fantastic maze of animal paths. Only flat-footed bonobos could be responsible for maintaining such a complex web of well-trodden trails. The tracks were strewn with recent feeding remains, pieces of the bokombe canes plucked by bonobos and discarded after the apes had eaten the pithy sections. These canes grew in thickets wherever the canopy was thin and sufficient light broke through to the lower forest levels. Every now and again we came across an odd piece of discarded chewed cane. It looked as though bonobos simply plucked them *en passant* and ate them as they travelled, rather as we might eat a lollipop or a bar of candy. Another plant eaten by the bonobos was lolongote. The apes had tackled this in a different way, unsheathing and eating the spaghetti-like curled white leaf shoots. This is a food that is well known to the human population of the Zaire basin and I had already enjoyed such forest spaghetti whilst I had been staying in Père Paul's Boende mission.

We passed several trees in which there were old bonobo nests but still no sign of the apes themselves. The nests were neat, springy platforms of interwoven branches, indistinguishable to my eyes from the nests of the common chimps I had known at Gombe, although they were built in much larger trees at about twice the height from the ground. Noel pointed to a red plastic ribbon tied round the trunk of a tree. This marked it as having been used by bonobos a few weeks earlier but there were no signs of apes there now and we carried on deeper into the forest.

About two hours from camp we came to another ribbon-marked tree. This was more recent and was in fact the last place where Alison had sighted bonobos. Nearby were several green, fresh nests and a large silver-trunked linkoko tree, still covered in a good crop of round yellow fruit, ripening to scarlet. Two small black hornbills were feeding on the sticky harvest but they had to spend more time and effort on scraping their beaks noisily along rough branches to remove the sticky latex than on picking and eating the fruit. The fresh bonobo nests and the ripening fruit were hopeful signs that bonobos were still in the vicinity, but we drew a blank when we searched the surrounding area.

We met many large troops of monkeys, mixed bands of black mangabeys, mona monkeys and coppertails. I had never come across such a mixture in Asia, where different monkey species living in the same areas of forest tend to exploit quite different canopy strata.

A strange, loud crunching sound drew us off the path to investigate. Two bushpigs, bright orange in colour, rushed off at our approach. They had been feasting on hard nuts. Benjamin sliced open a nut with his panga and we tried the rich kernels, which were certainly very delicious. After this brief snack we continued our search. Benjamin led the way, followed by Noel, and I brought up the rear. We passed under a large tree, its branches obscured by the dense foliage. Benjamin and Noel were already moving away when a small piece of fruit dropped at my feet. I glanced up and noticed a slight movement. I whistled to the others and they drew back under the cover of a convenient bush to get a better view. A small black bonobo climbed out on to a limb, collected a couple of fruits, then moved back into the thick foliage again. I was thrilled to have seen one of these elusive apes at last.

Now that I knew where to look I realized that there were several dark inactive shapes at the top of the tree, but it was not easy to tell how many bonobos were resting there. It looked as though the apes had made day-nests for sleeping. Only the juvenile was active, feeding and playing on his own, sometimes directly above me, sometimes on the far side of the tree where Noel and Benjamin were stationed. Suddenly the juvenile climbed right out of the tree and, using the network of branches, crossed quickly and easily above me into another fruiting tree. He was now invisible and absolutely quiet and it was not until he dropped some stones from his new position that I was able to locate him. Perhaps the new fruit was not to his liking for he did not stay long, but crossed back to the first tree as silently as he had come. Now there were stirrings among the other animals too.

The bonobos fed busily. My juvenile and an infant, about two years old, started playing together. The juvenile lay on his back with a grinning play face and tickled and juggled the baby who was clambering all over him. Round and round the tree they chased, and still playing their game of tag climbed back to where two adults were still feeding. I could see the juvenile clearly through my binoculars, a black face with pink lips, curious cheek tufts of hair rather like a siamang, spindly long arms and legs and a plump pot-belly. His little pink penis confirmed my impression of his sex, as he gazed absently over the damp forest around us.

Suddenly the whole party started to move. A large female led the way. She crossed the liana bridge over the path, paused briefly in a vine-draped tree then started to descend towards the ground. The infant clung tightly to her side while the juvenile followed close behind and an adult male brought up the rear. The adult female nearly reached the ground before she spotted us. She climbed hurriedly back up the vine, emitting loud piercing cries, furious and excited. The other animals started screaming with her like a chorus of shrieking herring gulls. The female moved hurriedly away through the treetops and I was enormously impressed by her speed and agility. I reckoned she was about two-thirds the size of a chimpanzee and she certainly showed much greater assurance in the treetops than the heavier Gombe chimps, who had always moved in a very careful manner when more than a few feet off the ground. At a safe distance she paused for the other animals to catch up. The juvenile was only a short way behind but the male was well back, shrieking at Noel and Benjamin. I hurried after the mother bonobo to get in front of her. Seeing this manoeuvre she turned to dash off at a new angle but when I again countered her movement from the ground she stopped and stared at me. Her infant scrambled round to peer at me curiously, but his mother maintained a haughty disdain. She started to go back the way she had come. I could have kept her in view for much longer but there seemed little point. The bonobos were obviously very disturbed at our presence and only wanted to get away. All we could see of them now was four frightened animals trying to outmanoeuvre us in the treetops and all the while they were shrilling their strange high-pitched calls of alarm at the tops of their voices. I walked back to Benjamin and Noel and together we watched the bonobos moving away through the leafy crowns. A hundred yards ahead of us they slipped down to the ground and vanished silently and invisibly along one of their well-worn trails.

I was very pleased with our encounter. We had watched them for over an hour before they had seen us and had enjoyed a good view of them even after that. Even their behaviour after they had seen us was instructive, for it was very different from that of common chimpanzees. At the sight of men, common chimpanzees will drop immediately from their trees and run off along the ground. Such behaviour is typical of other terrestrial forest primates such as baboons and pig-tailed macaques. The bonobos, however, had felt safer high in the canopy while we were close and had only dared descend to the ground when they were sure we were some way off.

During the next few days I was able to see bonobos on three more occasions. Early one morning we found a family group of a male, a female and a juvenile feeding in the fruiting linkoko tree. Unfortunately the animals soon saw us. They climbed swiftly to the ground and vanished without a sound. Another day we met a party of bonobos travelling along the ground but saw only a black flash as the leading animal turned and fled. My last encounter was with a lone adolescent bonobo that seemed to have been frightened by a snake. We heard her alarm calls from a distance and rushed up to find her shaking vines and displaying angrily at the ground beneath her perch. This strange performance lasted for several minutes but when she finally noticed her audience she redirected a brief display towards us, then climbed down and melted away into the forest.

In addition to these few exciting glimpses of the bonobos themselves I was able to learn quite a lot more about their habits from their nests and feeding sites. There was, for instance, a clear pattern to the distribution of bonobo nests. They were not built just anywhere in the forest; instead we found large clusters of them around major fruit trees. There were twelve nests around a loenza tree where the Badrians had seen bonobos feeding two months earlier. We found another eighteen nests around a linkoko tree that had already cropped, and we could see for ourselves how the number of nests around the large linkoko tree that was still in fruit gradually increased. These clusters of nests suggested that bonobos regularly visit the best fruit tree in their vicinity for the last meal of the day and nest close to that tree. They probably eat their first meal of the morning at the same tree before moving off for the day's forage. We had found bonobos resting in the high canopy at midday, but it was clear from the numerous feeding signs along the trails that they also spent much of their active time travelling and feeding on the ground.

Apart from the bokombe canes and lolongote shoots, bonobos eat a wide variety of animal foods collected on the ground. From examining bonobo dung, Noel and Alison had found that the apes had eaten termites and meliponid bees, giant millepedes, a snake and small mammals, including a shrew. Although there was no evidence that the bonobo's diet included anything as startling as the bush pigs and baboons sometimes eaten by the larger Gombe chimpanzees, it was exciting to learn that a forest ape eats small animal prey. The Badrians also found suspicious holes in the ground and soiled sticks were found nearby. It certainly looks as though bonobos are tool users like forest chimpanzees and use digging sticks for opening termite mounds.

The local Africans recount tales of bonobos catching fish from muddy forest pools and in one locality describe at length how bonobos dig for truffles. Some of these native stories about the small black apes are even more imaginative. There is a common termite in the forest which makes curious cone-tipped earth columns, a foot or two high but only a few inches in diameter. Sometimes these mounds are found lying broken in the middle of a bonobo trail. Occasionally the pieces form a square or triangular pattern with sticks laid between them. Probably these are the work of bonobos in search of termites but to the Africans this is sure evidence that the bonobos have tried to cook a squirrel. The bonobos, so the story goes, have seen

Africans making forest fires to cook small game and they copy them by building a ring of stones, or pieces of termite mound, with sticks in the centre on which to cook their prey. Unfortunately the bonobos don't know how to make fire for their food to cook and they eventually get fed up waiting and take the game away to eat it raw.

Sometimes large broken branches are found dragged along forest trails. We had found just such a branch on our first day out and these are not uncommon, particularly after rain. Probably they have been dragged along by bonobos in running displays like the behaviour commonly shown by Gombe chimpanzees, especially during strong rain. The Africans, however, have a more charming explanation. According to them, on wet mornings the male bonobo drags a branch along the path to knock the water off the adjacent vegetation so that his wife and children following behind do not get so wet. Such gentlemanly behaviour seems so alien to the Zairoise's own attitude towards his womenfolk that I am forced to believe that this explanation must be based on accurate, though misinterpreted, observation and that adult male bonobos have been seen performing dragging displays.

Even from my limited experiences with bonobos it is possible to make some comparisons with common chimpanzees.[2] On purely anatomical grounds the smaller, more Gracile form of the bonobo must be better adapted for travel in high canopy than his Robust cousin. I had been amazed by the agility of the bonobos; their treetop grace and sure-footedness were very different from the rather cautious and deliberate movements of the Gombe chimpanzees. In the rainforest the bonobos find most of their fruit in the high canopy and whenever they encounter terrestrial predators or have to cross swampy, flooded forest they also have to take to the treetops.

Bonobo social groups are also somewhat different from those of chimpanzees. The only large group of bonobos seen by the Badrians during nine months of fieldwork included many males and three females in full pink oestrus and it seems reasonable to suppose that these animals were together for social sex rather than normal foraging behaviour. All the other parties they saw, like those I witnessed, were very small. Many of these small parties contained both an adult male and an adult female. Moreover these groupings seemed more stable than those of chimpanzees, for Noel and Alison were able to recognize the same individuals travelling together after several weeks. In the Gombe chimp community the members of foraging parties were constantly changing, but sub-groups fell into three main categories: all-male groups, female family groups and rather larger mixed groups of both sexes. The small mixed groups of bonobos on the other hand seem more like the Gombe consortships where a single male travels around with a single female and her attendant young. Among chimpanzees such consortships are relatively uncommon, but the bonobo situation seems like a move in the gibbon direction of maintaining a permanent small family breeding unit. Chimpanzees form large social aggregations for grooming, playing and resting, with different sub-groups keeping regularly in touch with loud calling and displays. The community element in bonobo life is far less evident. The different foraging parties did not give loud calls nor did they show such ritualized and excited greetings as chimpanzees do when they met at fruit trees.

All of these differences between bonobos and chimpanzees can be explained by the fact that bonobos have become specialized for living entirely in a rainforest habitat. Chimpanzees do occur in equally dense rainforest in some parts of their range north of the Zaire river and in West Africa, but they are also found in such diverse habitats as their preferred parkland savannah, dry scrub, high altitude montane forest and deciduous woodland. As chimpanzees are wide-ranging and sexually promiscuous animals, there must be very fast gene flow right across their geographic distribution. Consequently they have not evolved to suit any one particular habitat type. Instead we find very much the same Robust adaptable chimpanzee over its entire range.

It is clear that the Zaire river has proved a sufficient barrier to genetically isolate the bonobo long enough for it to diverge from the common chimpanzee to the north, but who gave rise to whom? If the bonobo is really to be regarded as closer to the gorilla-chimp ancestor, or closer to the man-chimp ancestor, than the common chimpanzee, we have to imagine the chimpanzee as evolving from a small bonobo-like forest form. Everything I have learned of the bonobo suggests the complete reverse. To me the bonobo looks like a very recent offshoot of the common chimpanzee. The retention of the juvenile chimp body proportions, skull shape and even white tail tufts in adult bonobos looks like the result of neoteny, or the suppression and retardation of one set of physical developments, in this case body growth, whilst the onset of sexual maturity is hastened. The process results in sexually mature individuals with juvenile bodies and is one of the easiest and fastest ways that natural selection can produce smaller body size.

There is another reason for viewing the bonobo as derived from the chimp rather than vice-versa – the fact that the bonobo occurs south of the Zaire river. Most of the primates that occur north of the Zaire river have either failed to cross this broad waterway (e.g., all the baboons, drill and mandrills) or are represented by distinct rainforest forms on the southern side. In each case, the direction of geographical spread has been from the north and there is no reason why the bonobo should be an exception to this rule. The distribution of suitable chimpanzee habitat must have shrunk and expanded many times during the geological changes of the African climate. It would only require a southward extension of the rainforest by about three degrees of latitude to bring the distributions of chimpanzee and bonobo back into contact with each other in the Katanga rift area. This is the route by which the ancestors of the bonobo probably first reached the Congo river basin, and there is indeed evidence that these apes were once found this far south. The very first ape ever to reach Europe in the seventeenth century was described in detail by Dutch physician Nicholaas Tulp.[3] The ape was brought from Angola and from Tulp's description and illustration it was clearly a bonobo.[4] Bonobos survived in Angola until the end of the nineteenth century. Angola is far south of the present distribution of bonobos and is on the wrong side of the Kasai river. To have got there in the first place we must infer that bonobos once roamed even further south around the headwaters of the Kasai river, level with Katanga.

Several aspects of the bonobo's behaviour indicate that it has not always been a forest animal but has at one time enjoyed an important open-country, terrestrial ancestry. The bonobo exhibits well-developed, chimp-like knuckle walking. This is an effective and cheap way of covering long distances, ideal for an open-country animal but not so suitable for a small ape in thick rainforest where leopards and other carnivores may lie in wait. The quadrupedal mating postures of bonobos and the pink bottoms of oestrus females are also common features in terrestrial open-country primates, but they are not usually seen in rainforest canopy dwellers. The alarm behaviour of the bonobo is also ambiguous. The bonobo is afraid to descend from the trees when there are potential predators at hand and yet it obviously wants to flee along the ground as do all other terrestrial primates, including the common chimpanzees. All these features are exactly what we would expect from a recent chimp offshoot, but they are all rather out of place in the rainforest. If the bonobo had really enjoyed a long and uninterrupted evolution in the rainforest we would have expected it to be less ground-dependent and more like the arboreal Asian apes.

Those ways in which bonobos do differ from common chimpanzees, their smaller size, more stable male-female foraging partnerships, more ventral female genitalia for easier arboreal mating, and increasingly arboreal flight behaviour, all appear to be very recent changes improving the bonobo's efficiency in rainforest conditions. But what convinces me most of all that the bonobo is a comparative newcomer to the rainforest is the animal's lack of a locator call. So far as I know the bonobo is unique among forest primates in having no spacing, locator or territorial call. Surely if these apes had lived in the rainforest for long they would have developed territorial behaviour of some sort between different individual males. The only reason I can find for the lack of such territoriality in bonobos is that they were, until recently, co-operative community-living chimps.

For me the bonobo is a fascinating animal, not just because it is still something of a mystery and a challenge, but because it adds interesting light on the whole chimpanzee complex. In returning to a more traditional ape habitat the bonobo may actually have reverted to, or paralleled, some aspects of the ancestral forest apes. But the bonobo should be viewed as a recent rainforest offshoot of the woodland chimpanzee and as such is more distant from man than the chimpanzee itself.

CHAPTER 6

Orang-utans

BORNEO AND SUMATRA 1968-71

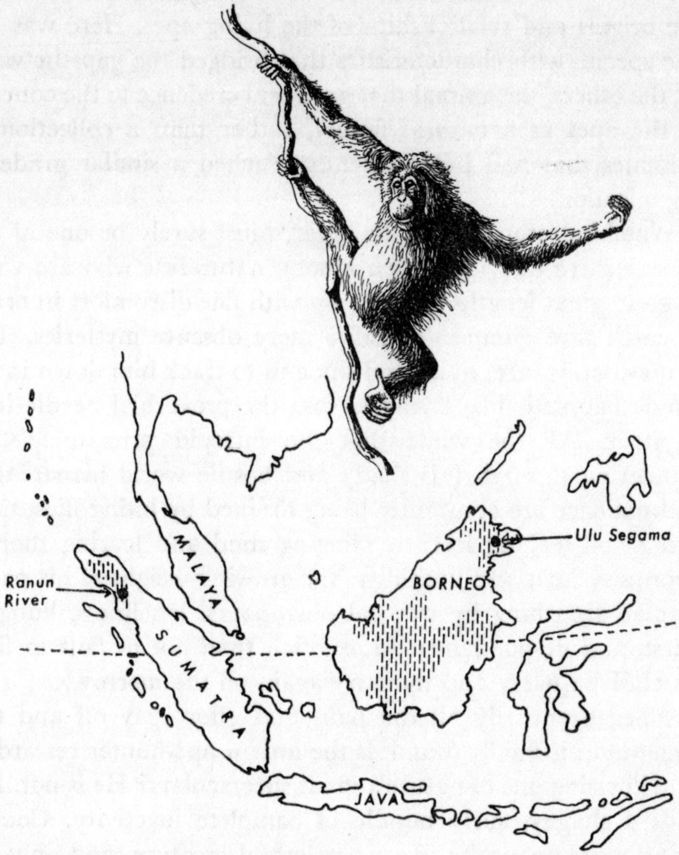

I tackled the orang-utan, or Asian 'man of the forest', shortly after my experiences with the Gombe chimpanzees, and from the start I approached the animal as though it were some aberrant type of chimp. What I discovered, however, was something very different – a completely new variant on the apely theme: an unlikely hotchpotch of spare parts, with a chimpanzee's body but a gibbon's long arms, the gorilla's sexual dimorphism but its own uniquely solitary habits; all within the framework of a strangely man-like society.

I soon realized that in the orang-utan lay the best clue as to the origins and relationships of the living apes. Here was the one species with characteristics that bridged the gaps between all the others, the animal that gave real credence to the concept of the apes as a natural family, rather than a collection of primates that had independently reached a similar grade of organization.

Watching orang-utans, however, must surely be one of the most bizarre activities, even among naturalists who are wont to go to great lengths and put up with fair discomfort in order to catch rare glimpses of Life's more obscure mysteries. The orang-utan is rare, quiet and shy and to track him down in his jungle haunts is like trying to find the proverbial needle in a haystack. All the while that the intrepid primatologist is wandering through this shady and hostile world his strength and patience are constantly being drained by biting flies, ticks and blood-sucking leeches, clinging mud and tearing thorns, oppressive heat and humidity, the growing weight of his pack, regular drenching by tropical downpours, loneliness, hunger, thirst, and despondency. More often than not he fails to find his elusive quarry and must try again on the morrow.

When eventually all the pain and effort pay off and the orang-utan is finally found, is the ardent ape-hunter rewarded by witnessing one of nature's great spectaculars? He is not. He finds a shaggy, surly bundle of complete inactivity. Cousin orang turns out to be the most slothful creature, and only by constantly reminding himself how lucky he is to see such a rare and exotic animal on such intimate terms can the primatologist fight off the urge to head straight back home. He must steel himself to remain in mutual contemplation with his subject for the long tedious hours of observation that are necessary to achieve those rare moments of interest and novelty that the orang offers as the only medals for persistence in this game.

Only months after discovering how many notebook entries it takes to establish that an orang has not moved in the past five hours, and long after he is already a world expert on the behaviour of leeches and ants and the decay rates of rainforest leaf litter, does the ape-watcher glean enough interesting incidents of orang-utan activity to form a clear idea of the red ape's society. Hyper-active gibbons and chimps give the field-worker no chance for mind-wandering or boredom and even the less dynamic gorillas travel in such large groups that there is usually one animal doing something of interest. There can be no doubt that the solitary, slow-moving orang-utans clear the board when it comes to tedium; yet these sensitive creatures remain my favourites of all the apes and it is to this species that I have devoted most of my time.[1]

My introduction to the ways of the orang-utan came amidst the uninhabited forests of the Ulu Segama Reserve in Sabah, Borneo's north-east corner. Here two Dusun boatmen maintained my camp beside the swirling river, collecting firewood, hunting wild pig and catching fish, while I searched the hills and stream valleys for the red apes. During my first four months I met an encouragingly large number of these strange, solitary apes and, spurred on by this success, I managed to raise sufficient funds to return for a further year of study.

As my knowledge of the forest and my quarry improved I was able to overcome many of the problems in locating the apes. The secret, I learned, was to travel alone, very quietly and slowly, pausing often to sit on fallen logs and listen to the myriad sounds of the surrounding forest. I learned to pick out tiny noises filtering through the hum of insects and chirping birds that meant nothing in themselves but gradually accumulated to give me a picture of the invisible world hidden behind the enveloping greenery. Quite suddenly I would become aware of a troop of macaques or a feeding orang-utan even though they might be a hundred yards or more away. Sometimes it was a smell lingering on the morning air that led me to an orang, or a diligent search round a fresh green sleeping-nest. Sometimes I struck lucky by simply visiting the places where I knew there was a good tree in fruit, or I might track down a big male half a mile away by following a compass bearing taken on one of his infrequent, aggressive long-calls.

Once I had found my orang-utans it was quite easy to keep in touch with them because they were such slow movers, covering only about five hundred yards in the course of a whole day. Orang-utans are not as fleet-footed as chimps or bonobos. Their hook-shaped feet are specifically designed for clinging on to branches, and not very suitable for ground walking. Moreover, unlike the more terrestrial African apes, the orang-utan does not have calloused knuckles to protect his hands on rough terrain. Apart from these anatomical drawbacks, the orang's general slowness and solitary habits make him particularly vulnerable to terrestrial predators and he feels much safer in the trees. Orangs often go to ridiculous and laborious lengths to find an arboreal route rather than visit the forest floor. Only the very large males, who may weigh as much as two hundred pounds, make regular use of ground travel. Indeed, tree-top locomotion becomes both difficult and hazardous for such monsters and their enormous strength is enough to frighten off any marauding cloud leopards.

Up in the trees the orang is amazingly effective considering its large bulk. It does not travel much through the upper canopy layers, like the more agile bonobos and lesser apes, but prefers to use the lower and middle storeys which are so well packed with small trees in the Asian rainforest. The orang progresses slowly and noisily, clambering across from one crown to the next and often swinging small trees to and fro to reach across a gap. Nevertheless it was easy enough to lose an orang-utan in the forest unless I stayed very close. Once out of sight the animal was surprisingly difficult to relocate; its stationary red bulk blended very well with the greens and browns of the surrounding vegetation.

To maintain contact with a subject over several days I would have to be back at the orang's night-nest before he left in the early morning. If the nest was far from my camp I had no alternative but to spend the night in the forest. I built several night shelters in useful locations throughout my study area and stocked these with tinned food, but I often found it easier to simply wrap myself up in a large sheet of polythene and settle down between the convenient buttress roots of a large fig tree. Sometimes I was away from camp for more than a week, but my Dusun helpers became used to my irregular habits, confident that I could look after myself.

Gradually my observations began to add up. I began to understand the orang-utan's moods and rhythms, accept his easy-going pace of life, see pattern in his apparently formless society. From encounters with animals of all ages I was able to build up a fair picture of the red ape's life history and I began to recognize the individual characters of some of my more regular contacts.

Redbeard and Harold were rival adult males and the main ridge across my study area seemed to be a disputed region of range-overlap. Whenever Redbeard was on Middle Ridge he seized the opportunity to direct a barrage of loud aggressive long-calls westwards, giving at least one impressive call series a day and sometimes two, or even three. Whenever Harold was on Middle Ridge, he would call towards his rival's eastern range, but the two animals made sure they were never on the ridge at the same time and thus carefully avoided a show-down.

The loud long-calls are a means by which the adult males can inform one another of their whereabouts, so that they can space themselves out and avoid meeting face to face. Occasionally the system breaks down and one animal does blunder into another. Twice I have seen such confrontations: they were brief, violent affairs. The orang-utans glared at each other and then one male backed down by turning away and fleeing, pursued by the displaying victor. Their great face-enlarging cheek flanges, high fatty crowns, extra-long beards, flowing body hair and overall large size, made the displaying adult males a very impressive and intimidating sight. Overawed by the spectacle of such large and frightening opponents, the faint-at-heart soon took to their heels so that the confrontations were usually settled without resort to physical violence. Sometimes, however, the trespasser refused to back down and the male orang-utans fought, grabbing and biting at their rival's hands and face.[2] Most of the adult males had bite scars on their cheek flanges and a few, like Harold, had stiff fingers, which were probably the result of injuries incurred in status fights.

Just as I could recognize the deep bubbling tone of Harold's calls so I am sure that different male orang-utans could recognize their neighbours' calls individually. Sometimes, however, they made mistakes, as Redbeard did one day in May.

At 8.50 I heard Redbeard calling somewhere up near the hill I called the Picnic Spot. I set out to look for him. By 9.30 I was able to pin-point his position from the sounds of falling débris. Redbeard was chewing the bark off vine stems, sucking out the sweet juice, then spitting away the discarded fibrous pellets, like an old man chewing on his tobacco. Suddenly at 9.45 I heard another long-call some distance to the north-west. Redbeard, too, heard the call and responded by climbing to the ground and marching towards me. He looked menacingly purposeful and I hurried out of his way. I don't know whether he even saw me as he marched determinedly past towards the calling challenger.

I followed cautiously for about three hundred yards and watched him climb up into a small tree to give a long-call of his own directed towards the invisible protagonist. Redbeard looked magnificent, his great head raised to the sky and his throat pouch greatly inflated as he gave twenty deep, roaring groans. The whole tree swayed vigorously through the display then, momentarily satisfied, he trailed off the call in a bubbling finale and started to feed on vine shoots.

At 10.50 Redbeard's persistent antagonist announced his new position, almost due north of us and rather closer than before, by a tremendous series of long-calls. Redbeard was furious and replied with an even more impressive display, an incredible long-call accompanied by violent tree-shaking. Unappeased he set off on foot once more towards the other caller. He eventually climbed into a large tree on the northernmost ridge of my study area. All remained quiet till at 11.45 there was yet another call ahead of us. It was a double call with two climaxes, a trick used by Raymond. Immediately after the call there was a great crash of falling branches from the same direction, perhaps a natural fall but more likely the result of the displaying male breaking off a dead limb. The crash was followed by yet another impressive long-call.

Surely Redbeard must have known whom he was dealing with now. He made no further reply but slipped quietly down to the ground and hurried away to the east. Raymond called another four times over the next four days, but Redbeard kept a very low profile and it was several days before I heard him dare to call again.

Raymond was a very large male, quite at home on either side of Middle Ridge and he seemed to have an emasculating effect on Harold as well as Redbeard. Harold also kept quiet for a week after hearing a single call from Raymond, although he had been calling regularly until the dominant animal arrived. But although Raymond put fear into the hearts of his orang-utan competitors, he was terrified of me. Whenever he saw me he hid and sulked in the top of a tree, refusing to move for hours on end. Although he was an irregular visitor to my study area I still regarded him as a resident, but there were many other males who also called but only appeared once during my whole study and seemed therefore to be leading quite nomadic lives, greatly upsetting the resident male hierarchy whenever they passed through.

There was no love lost between the different male orang-utans as their calling, avoidance behaviour and aggressive displays clearly demonstrated. Nor was it uncommon to see adult female orang-utans hide in the top of trees to avoid being seen by such bad-tempered males when they passed them in the forest. Such shyness was well-founded, for I witnessed several quite brutal attacks by males on females. Sometimes these were sexual assaults but often they looked like mere aggressive spite. Usually, however, the two sexes went their own sweet way and showed complete disdain and lack of interest in each other, even when they met casually at good fruit trees.

Adult females were no more friendly among themselves. Each female travelled separately, accompanied only by one or two dependent young. Different animals ranged over the same parts of the forest, so would meet at food trees, but showed no greetings, no anxiety, no interest. Different females with their respective youngsters would arrive, feed and leave the shared food tree in a totally individual fashion, although as many as eight different animals might actually be sharing the food tree at the same time.

It was only among juvenile orang-utans that I saw real signs of social tendencies. Like other young primates, juvenile orangs are playful and inquisitive. They become bored with their mothers' company and yearn for playmates with whom to share their fun. When Rita and her three-year-old son, Roy, met up with Joel and her two youngsters the two mothers ignored each other. Rita rested aloof whilst Joel cradled her suckling baby. The two juveniles, however, dashed to greet each other and joined in a glorious half-hour of exuberant rough and tumble. With curled wrists Roy and Joey struck out at each other, chased madly round the crown of the small tree, swung excitedly on vines, grappled and bit, their huffy laughter and grimacing play-faces emphasizing how much they were enjoying themselves. It was Rita who broke the party up; without warning she set off down the valley. Roy was too engrossed in his games to notice his mother's departure. Suddenly he realized that she was gone and he rushed off down the slope uttering harsh and angry screams until he caught up with her again. Joey stayed behind with his own family but now had to content himself in gentle, unmatched games with his baby sister, while mother Joel enjoyed a midday snooze in a loosely-built day-nest.

If all youngsters have such a lonely, friendless existence it is easy to understand how adult orang-utans come to lead such solitary lives. In infancy and juvenilehood they have few opportunities to establish the play-peer relationships that are so important for the proper social development of chimps, gorillas and indeed most other higher primate species. And so, unused to making friendly social relationships, orangs grow up to be shy, solitary, anti-social adults.

The young orang-utan grows up in a world almost entirely filled by its mother. Only a few lucky infants enjoy an elder sibling like Joey. Most infants are born at three- or four-year intervals, so that the older offspring is leaving the mother just when the next is starting to want a playmate. So the playmate has to be mum, but she is usually busy feeding or resting and never gets as involved in games as young orangs would like.

The arrival of a new infant is a major threat in the life of its elder sibling. Suddenly the sole apple of the mother's eye is relegated to number two. Suddenly he is no longer welcome in her warm night-nest and must make his own little bed a few feet above her. Suddenly he is no longer entitled to a ride on her side over tricky tree routes, no longer able to suck her nipple for reassurance, no longer able to command his mother's attention at the faintest whimper. Overnight the youngster is forced into a juvenile life-style and his privileged and spoiled position is usurped by the tiny, spidery newcomer.

The juvenile may stay with his mother for a year or more after the new baby is born, but all the time he is looking for something else in life. Neither his mother nor his younger sibling satisfy his growing social needs. Just such a juvenile lived in the forest to the south of the Segama river, an area I explored during my first trip to Borneo. I called the animal Monty and when I met him he was alone. He was about four years old and at first I thought he must be an orphan.

Monty drank thirstily from a deep tree-hole, scooping up water quickly with a cupped hand and sucking slurpily as the water spilled out of this imperfect vessel. He swung off with enviable ease through the tree-tops and eventually climbed into the crown of a large oak tree. Only then did I realize that he was not alone. A large dark, chocolate-coloured female and a small infant, his mother and brother, were already sampling the hard acorns.

I saw quite a lot of Monty and I was struck by his independence. He had no qualms about wandering off on his own for several hours at a time, but every night he would rejoin his family to indulge in a short play session with the baby whilst mother made her night-nest. As darkness fell baby Mitz climbed down to share mum's bed and Monty started his own nest-building just overhead.

One day Monty took a fancy to a passing sub-adult male and tagged after him. Poor Arnie was not at all flattered by such juvenile attention but could not shake off this unwanted companion till the evening. Monty had better luck the next day when he met a mother with a juvenile female of his own size. For over an hour the two juveniles enjoyed a boisterous game of tree-top tag, and I watched enthralled as the two furry bundles grappled and chased, nibbling each other's hands and feet and swinging to and fro Tarzan-fashion among the hanging lianas.

Monty's stage of semi-dependence precedes the complete split which occurs between mother and offspring at the adolescent stage. Males seemed to leave their mothers earlier than females but the eventual break-up is just as inevitable, and I observed several adolescent females who occasionally rejoined their mothers for a day or two at a time but normally travelled alone. Sometimes adolescents that had left their mother tagged briefly on to other family groups, or partnered other lonely adolescents or even young adult males, but as adulthood drew on they preferred complete independence.

But why are adult orang-utans so solitary? Why do Joel and Rita lead such separate lives rather than travel together with united families? The answer lies in the distribution of fruit foods in the Asian rainforest. Fruit trees of the same species do not stand in clumps but are widely scattered. Thus there are large distances between trees that are in fruit at the same time, but the orang-utan is slow-moving and able to search only a small area each day. Two families travelling together would find no more fruit trees than one family alone but the food they found would have to be stretched to feed twice as many mouths. The price for group-living would be high. Animals would either have to make do with smaller proportions of the rare fruits in their diet or else travel further than they could afford in search of extra fruit. However sociable juvenile orang-utans might like to be, they soon discover that they cannot afford social luxuries. Except under exceptional conditions of fruit abundance they must live in as small a group as possible. Mothers are accompanied only by their dependent young and the young leave their mothers as soon as they are self-sufficient. Males usually travel alone and only consort with females for mating purposes as rarely as the slow reproductive needs of this long-lived species requires.

The energy costs of arboreal travel for such a large animal as an orang-utan are so enormous that it is amazing it can make a living as a fruit-eater at all. Yet despite travelling less than one quarter as far as a gibbon does, and having to find approximately six times as much food each day, the orang-utan manages to obtain a larger proportion of fruit in its diet. The orang-utan shows a simply uncanny ability to locate fruit, and his secret is superior brainpower.

Psychologists working in laboratories have been very surprised by the intellectual abilities of captive apes, intellectual abilities that seem way beyond anything they may need in the wild. Intelligence is often thought to be a reflection of the social complexity of an animal's way of life, but the intricacies of chimp and gorilla society are certainly no more complex than those found among the far less brainy baboons. And there can hardly be anything less complicated than the daily travel, feed and sleep routine of the solitary orang-utan. The truth is that ape intelligence was evolved in response to the difficulty of locating fruit in tropical rainforests.

The jungles of Borneo contain something like three thousand different species of tree in addition to a multitude of epiphytes and herbs. Within the home range of a single orang-utan there may be as many as five hundred types of tree, half of which will provide some sort of food at one time or another. The problem is to know what food, and particularly rich fruit, will be available when and where. The matter is further complicated by the fact that very few of the forest trees are annual in their flowering and fruiting cycles. Some, like the figs, fruit regularly every few months but many other species may fruit only once every two, three or more years. As a result the distribution of fruit varies throughout the year and no two seasons are ever quite the same. Faced with the same problem the gibbon patrols his small territory regularly, often daily, to see what is ripening where. He needs to make few predictions. The orang-utan, however, marches slowly through parts of the forest he has not visited for weeks or months, yet he works steadily and directly from one good food source to another.

I was fascinated by the orang-utan's ability to turn up in the right place at the right time. Often I quartered the forest all day looking for good fruit trees and did not find one. If I did it was usually either already occupied by an orang-utan or one soon arrived. The best way to find orang-utans was to visit the trees that were in fruit and the best way to find fruit trees was to follow an orang-utan.

To forage so efficiently when fruit is always rare and scattered, orang-utans must have an incredibly accurate memory for both the location of many different trees and also their past fruiting histories. From this knowledge the ape can predict with reasonable accuracy where there is food and what is his best route to reach the harvest.

Sometimes the orang-utan's attention is drawn to a fruiting tree by the noisy behaviour of feeding hornbills, pigeons or monkeys. They can also make use of the intimate and up-to-date local knowledge of gibbons. On one occasion I saw a group of three orang-utans rouse from a morning rest at the sight of a family of gibbons swinging purposefully past. The orangs hastened after the smaller apes and finally caught up with them at a small tree laden with sweet plums. The unfortunate gibbons who had led the way were not allowed to share in the prize and had to make way for their larger cousins, who stripped the tree.

I am sure that the orangs can remember where they have seen trees of their favourite fruit in flower, or bearing immature fruit, and are able to calculate when it will be worthwhile revisiting them. Moreover, although many rainforest trees fruit only very irregularly, trees of the same species fruit at the same time. Consequently, if he finds a fruiting shurud tree the orang knows that other shuruds will also be fruiting this month and he can draw on his remarkable memory for information about the location of those trees.

With regular favourites like wild durian fruit, orang-utans undoubtedly carry mental maps of the location of individual trees. During the durian season in July they simply travel from one tree to the next. Only about a quarter of the durian trees fruit in any year but there are always some to reward the canny orangs. Knowing that durians were so popular I had taken the trouble of mapping all the eighteen durian trees I had come across on my forest perambulations. When July arrived it turned out to be a poor crop but on a single day I followed the sub-adult male, Humphrey, on an impressively economic course that took in six of my trees. Even with my map I could not have made a shorter trip to visit so many of the trees!

It is quite clear that orang-utans do not amble aimlessly around their ranges taking pot luck. They move in a very purposeful and deliberate fashion. To plan such routes they must not only have a very good knowledge of what is where, but they must also have a good sense of direction, distance and travel time. It is no good setting off for a durian tree that you think will be laden with juicy fruit if it is two days' travel away and there is nothing to eat en route. The orang-utan must use his detailed knowledge of the range to plan the best route with stops for meals on the way to his ultimate goal. All this takes intelligence and a retentive memory but as a result the orang-utan is several times as efficient as a random fruit-seeker. For such a heavyweight frugivore this is the difference between survival and starvation.

The animals in my study area showed seasonal shifts in their ranging patterns so that in some months most of them were visiting the low-lying riverine forest while in others they withdrew to the foothills. Different areas of the forest were clearly seasonal in their productivity. At some times of the year there was plenty of fruit, at other times fruit was scarce. Obviously the number of resident animals that can survive in any locality is the number that area can support at the leanest time of year, but in more plentiful times the area will be able to support a few extra nomads. If food shortages or abundances were similarly timed over the whole forest there would be no opportunity for nomads to exist, but in the Segama Reserve different forest areas were clearly not in synchrony. Because of slight differences in topography neighbouring valleys supported quite different tree species; moreover the same species would fruit at different seasons in different valleys. Several species of fruit ripened a month earlier on the southern side of the Segama river than they did to the north. By simply moving to new areas as the fruit there developed, orang-utans could prolong the seasons of favourite foods, and some nomads could make a living in this way by travelling from one area of temporary abundance to another. Nomads must always be at a disadvantage when competing for food with residents since they have a poorer knowledge of local geography and are challenged by resident males, but nomadism is certainly better than starvation if you cannot find a permanent range of your own. Moreover by moving away to areas where the trees grow greener, nomads do not have to face the local periods of food shortage when fruit is scarce and resident animals have to get by on a diet of leaves and vine bark.

During June and July my study area became quite over-run by new visitors. I met thirty-one different orang-utans during that period. All the animals were travelling in the same direction, towards the east. In August, some nine miles to the east of us and close to my boatman's kampong, there was a much better durian harvest than ours. As if by magic the orang-utans arrived just in time to compete with the villagers for the valuable fruit, although at other times orang-utans were never seen in the vicinity. I can't help thinking that these were the same orang-utans that had passed through my own area earlier and I remain teased by the question – did those animals know that there was going to be a better durian crop there?

There can be no doubt that the Segama orang-utans were essentially solitary animals with widely overlapping individual ranges. Although some resident animals remained in small areas, just sufficient to accommodate all their needs and perhaps no more than a square mile in size, other animals covered much greater distances. The very loose nature of the orang-utan society enabled the population as a whole to show enormous flexibility in its spatial distribution which could vary intelligently to suit prevailing conditions. If food was concentrated then orang-utans would congregate to make best use of it, but when food was scarce or dispersed then the animals scattered to forage over a wider area. Such strategies are quite different from those employed by the Bornean gibbons which looked for fruit in these same forests, because the strict territorial system of the smaller apes prevents any such flexibility.

SUMATRA, 1971

Orang-utans today occur only on the islands of Borneo and Sumatra. The two populations have been separate for many thousand years and I was interested to see if and how the Sumatran orang-utan differed from his Bornean cousin. In 1971 I made a seven-month field trip to North Sumatra. During brief surveys in the Langkat and Gunung Leuser reserves I was able to watch orang-utans amidst fantastic scenery. Sharp limestone cliffs draped in green forest, tall waterfalls, rushing streams and umbrella-like tree ferns all added to Nature's artistry in shaping the most beautiful ape habitat I had ever imagined. Nevertheless the steep gradients made both reserves very difficult for field work and I ultimately settled on an area of lower-lying forest, well trampled with convenient elephant paths.

My new Ranun camp consisted of a single thatched hut in the forest but there was none of the wealth of fish and game that I had so enjoyed in Borneo. The standard of living was incredibly low and there was often no more than rice and chilli peppers for supper. Of orang-utans, however, there were plenty and I soon found myself filling up notebooks with the everyday dramas of their lives.

Two calling males, Co and Mo, proved to be the dominant characters in my new study area and competed with long-range displays in much the same way as Redbeard and Harold had done in Borneo. The calls of Sumatran males are less prolonged and less dramatic but obviously long-calls just the same. As in Borneo, other small foraging parties remained within earshot of their calling patriarchs but no one liked to get too close and animals hid in tree-tops if there was no other way of avoiding an encounter. A great deal more of Sumatran orang-utan society was similar to the picture I had witnessed in Sabah but there were some interesting differences, and I was glad I had changed islands.

Among my early acquaintances in Ranun were a trio I called Karl, Kate and baby Kim. Karl was a handsome sub-adult who was the constant and apparently devoted companion of the large, long-haired Kate and her golden infant. The trio stayed together for the entire period of my study, which was far longer than any such relationship had survived in Borneo. Moreover, Karl occasionally groomed Kate and played with the baby, behaviour quite foreign to any self-respecting Bornean orang-utan. Nor was this consortship in any way unusual. During my stay in Sumatra I met several other orang couples foraging together and some of these also remained together for several months.

There was more evidence of a close-knit society in Sumatra. When a huge strangling fig fruited on the far bank of the Ranun river it attracted fourteen different orang-utans over a five-day period and many of the animals fed together. These orangs not only tolerated one another but on two occasions I saw new arrivals approach settled diners and inspect them in a sort of greeting. Several times I saw more than one adult male feeding in the same tree but there was practically no status friction. One male did chase a fellow from the tree but both animals were feeding peacefully together again the following day. Such inter-male tolerance would have been inconceivable in the Ulu Segama. This is not to suggest that all was love and friendship. There was one very large male with whom no one would share a tree. Each time he arrived, with a spectacular branch-shaking display, the other diners left the tree in haste. On the whole, however, it was their tolerance of one another that struck me most.

Apart from these increased levels of co-operative behaviour the Sumatran orang-utans seemed to be less scattered than my old Bornean friends. The Ranun animals travelled around in the same tiny groups, but the groups seemed more clumped and better co-ordinated in their movements around their shared range. My study was too brief to confirm these impressions but I certainly felt that I was dealing with something more like a loosely-knit chimpanzee community than the totally independent ranging patterns of individual Bornean orang-utans.

There are several good reasons why Sumatran orang-utans should need to keep in closer contact and be more co-operative. They face more serious predators and fiercer primate competition than their Bornean counterparts. Tigers do not occur in Borneo but they have ample opportunity to take their toll of unwary Sumatran orangs when the latter are forced to descend from the safety of the canopy to cross rocky screes. It is in just this sort of vulnerable position that one benefits from neighbours, for more neighbours mean more scouts. Any orang-utan meeting a predator like a big cat will squeak and grunt or give loud lork calls that give fair warning to friends within earshot.

Sumatra is also one of the haunts of our old friend the siamang. Siamangs are only a quarter the weight of orangs but they are much faster, and travel in family groups. I have watched a family of four siamangs swoop like threatening furies on a young orang-utan feeding in a fig tree and they were only routed by his mother's furious display in defence of her youngster. On another occasion the same family of siamangs succeeded in driving a lone sub-adult male orang from a fig tree. The orang-utan, terrified by the sudden onslaught of these black imps, rushed to the end of his branch and leaped out of the tree to crash-land in the crown of a smaller sapling below. Only in great emergency does the cautious orang-utan throw himself into space. Obviously siamangs can be quite a threat to orangs and it was only when there were several orangs feeding together that their small black cousins had to keep to a respectful distance.

Overall I found the Sumatran orangs to be very similar to those in Borneo, but, while the scattered distribution of food in both countries favours solitary habits in orang-utans, the Sumatran animals face greater predation and competition problems than their Bornean counterparts. In consequence it has paid the Sumatran orang-utans to develop slightly greater social tolerance, co-operation and cohesion within their community.

The sex-life of orang-utans is bizarre to say the least.[3] Even an anti-social and solitary animal, however long-lived, has to breed sometimes, and as a pre-requisite for procreation males and females have to get together. Compared with the sexual excesses enjoyed by chimps, the orang-utan is very restrained. An observer of chimpanzee behaviour may witness a dozen or more copulations a day but the orang-utan *voyeur* is lucky to observe such behaviour once a month. However, though the orang-utan may not indulge frequently, when he does mate he makes it last. Orang-utan sex is a prolonged and erotic business, very different from the ten speedy thrusts of a male chimpanzee.

Orang-utans will mate in trees but occasionally they end up on the ground. They can perform in a weird variety of acrobatic suspensory postures with males mounting from in front, behind, above and below, usually clasping their partners around the thighs with hand-like feet and thrusting in a slow, deliberate tempo. Sexual encounters may be very aggressive with screaming, man-handled females struggling and fighting to be free. Alternatively they may be co-operative affairs and sometimes females even take the initiative. Sex may be the prerogative of the dominant-calling male but such patriarchs are quite unable to exert any sort of monopoly. They cannot keep watch over a thousandth of their range and opportunities for cuckoldry are immense. Sub-adult males court females in their own quiet manner, resident males expect receptive females to be lured by their splendid long-calls, whilst nomadic males *en passant* adopt their own rape tactics in the great competition for reproductive success.

Sub-adult males are a class of sexually mature, but not yet full-sized, animals. They are aged between ten and fifteen years, but have not yet developed the big cheek flanges or long hair of the full adults and do not dare to invoke attack by advertising themselves with premature long-calls. They are thus unable to call up potential mates and must actively search out members of the female sex with whom to consort. Full adults, on the other hand, have no hesitation in broadcasting their whereabouts and any female interested can thus choose with whom she will mate. In practice, consortships with full adult males are rarer than those with sub-adults and are generally briefer, lasting only a few hours or a couple of days, although they may be maintained intermittently over longer periods. Because of his great bulk it is in fact quite difficult for a full adult to stay close with an adult female. Big males want to travel on the ground while females prefer tree routes. Big males want to stay longer feeding at each good food site but the half-sized females prefer to visit more sites each day. The two sexes are just ecologically incompatible. Since only receptive females are attracted to the calling, however, the male enjoys a high level of success from his infrequent sexual adventures. The sub-adult male has to work much harder. He is often consorting with young, inexperienced females or females of reduced fecundity accompanied by small infants. Or he may meet a potentially fecund female at the wrong time of the month and then have to stay

Siamang

White-handed gibbon

Gibbon

Above and overleaf: Gorillas

Left, right and overleaf: Chimpanzees

Left: Bonobo
Below and overleaf: Orang-utans

with her for a longer period if he is to have much chance of impregnating his partner. He can, however, cope more easily with extended consortships than the big patriarch because his smaller size makes him more ecologically compatible with the female's daily feeding and travel routine. The courting subadult overcomes the female's shyness and suspicion by gentle tactics and playful contact, interspersed by undulating postural and hair-waving displays. Play may lead to sex, for female orang-utans can be receptive at all times of the month and do not go in for marked sexual swellings like oestrus chimpanzees.

I felt that female orang-utans were also exploiting this false sexual receptivity to retain male consorts whilst they still had tiny infants. A female like Kate could not possibly have been fecund so soon after the birth of Kim, yet she and Karl enjoyed long sessions of rather playful sex. Probably such male consorts served a protective function through the vulnerable mother-infant period. In Borneo, where predator and competitor threats were less severe, females raised their infants unaided and were not consorted by males, but in Sumatra almost all females with young infants were accompanied by a consorting male or an older offspring.

Very old males both in the wild and in captivity eventually become totally celibate. Indeed, we know these captive males to be impotent and it seems fair to assume this is also the case in the wild. Neither Harold nor Raymond were ever seen consorting, nor did they seem at all interested when apparently suitable females were available.

Shortly after Harold had given one of his spectacular calling performances he and I both heard the unmistakable sounds of approaching orang-utans. The newcomers were a large adult female accompanied by a three-year-old juvenile. She cannot have failed to have heard Harold's calls and from her direct approach it seemed obvious she was actively seeking his company, but Harold did not want to know. He slid down a liana to the forest floor and sneaked off quickly and without a sound. The female arrived to find no magnificent male orang-utan, merely an interested primatologist, but her haughty glance left me in no doubt that she did not for a moment credit me with the splendid performance that had drawn her to the ridgetop.

Nomadic males have no time for lengthy consortship with the resident females whose ranges they pass through. Sometimes females may be attracted to their calls but are probably shy of the total strangers. This does not mean that nomadic males must forgo their sexual pleasures. Several times I have seen them as they chase, catch and simply rape unwilling females they have met in the forest. Resident males may also rape nomadic females and I once saw Redbeard chase a strange female for over two hundred yards before finally catching her, dragging her out of her tree and raping the unfortunate screaming creature on the ground.

In its sexual behaviour the orang-utan makes a much better stepping-stone to the human condition than does the chimpanzee. The chimps' speedy performances and the obscene sexual swellings of oestrus females seem very foreign to our own species. Sexual behaviour in the orang-utan seems much more familiar. Orang sex is a mutually erotic and lasting performance, imaginatively varied in form, context and position. As in our own species, orang sex can be aggressively male initiated, mutually entered or female initiated. Females show month-round receptivity. Young couples indulge in rather playful courtships with the young males trying to impress their initially shy pick-ups with their physical splendour, and such flirtations sometimes lead to lasting sexual consortships.

In Sumatra orang-utan society shows a loose organization with individual animals all going their own independent ways within the same shared tracts of forest, a pattern not very dissimilar from life in a chimpanzee community in fact. Within this rather nebulous community the adult male orangs compete for rank and local status by means of long-range spacing calls, but the overall situation remains relatively stable and peaceful. Adult animals form consortships which sometimes develop into lasting protective pair-bonds, with males escorting small mother-family units. As in Borneo, the loose nature of the society gives the population sufficient flexibility to allow these intelligent animals to vary their distribution to suit changing social needs and patterns of food distribution.

I believe that this type of social organization makes the best common model for the derivation of all other ape societies, since fewer changes are needed to achieve these variants than are required from any other theoretical starting point. As we might expect, the Bornean orang system is almost identical except that there is even less community life and co-operation. Among gibbons the pair-bonds have become permanent and separate families have become territorial rather than range-sharing. In the gorilla system the whole community has become group-living, with animals centred around the dominant males. Such a way of life is made possible because of their terrestrial browsing niche. In chimpanzees the community is tighter-knit and more rigidly organized, with more even sharing of male responsibilities and reproductive rewards, but really the highly flexible system remains very similar to that of the Sumatran orang. The greater terrestriality of chimpanzees, which probably evolved from necessity in areas where the tree canopy was incomplete, permits greater individual mobility and hence greater cultural and social potential. In bonobo society we see a return towards the orang-utan condition brought about by the animals' return from open woodlands to a closed rainforest habitat. Bonobos are more arboreal than chimps, less social, less co-operative and show a greater tendency to live in bisexual family clusters. Even the human system of individual families developing within larger loose communities is very similar to the Sumatran orang-utan model.

It is not only in his social organization that the orang-utan helps us to understand the habits of the other apes. In his locomotion, too, he shows the greatest conservatism. Among the living apes we find arm-swinging brachiators (gibbon and siamang), quadrupedal knuckle-walkers (chimp, bonobo and gorilla), a striding human biped with long projectile-throwing arms and the orang-utan, who is an arm-dominated quadrumanual climber. Yet the orang-utan can resort easily enough to quadrupedal ground walking, is an accomplished biped and an excellent brachiator as well. The slow quadrumanual climber makes the best starting point from which to derive the diverse locomotor specializations seen in the other apes[4] and the orang-utan's very flexible shoulder joints also make it by far the best model from which the human shoulder pattern could have originated.

What then do we make of cousin orang? Socially, sexually and anatomically, he makes the best ape ancestor for man. Do we proclaim him as man's nearest living relative, the species closest to the common ancestor of man and ape? Or is it merely that while the other apes have specialized and diversified, the orang-utan has remained in a conservative niche in the ancestral habitat – the large-bodied, quadrumanual frugivore of the equatorial rainforest? Perhaps a perusal of the fossil and biochemical evidence will provide some of the answers, as we pry deeper into the past history of man and ape in our search for the ape within us.

Part Two

MEN-APES TO APE-MEN

CHAPTER 7

The Missing Links

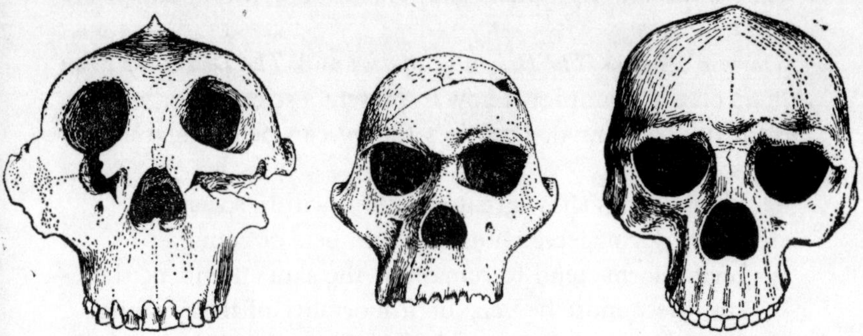

Man's fascination with his own nature is as old as man himself, but at the end of the last century this interest passed out of the sphere of philosophy into the realms of natural science and hard fact. Scientists had long realized that different living animals and plants resembled one another and could be conveniently classified into families and genera depending on their degree of similarity, just as one can classify books in a library or any other collection of objects. It was, however, the very firm view of Christian dogma that God had created each species independently and in the precise form that we now see it. Although traditional ideas were challenged by geological findings of fossil creatures quite unlike anything surviving on earth, and of depths of rock sediments that told of enormous lengths of time (periods far older than the six thousand years since the presumed date of Biblical Adam), without serious alternatives to the Creation story doubters could be temporarily silenced by the idea of successive divine catastrophes followed by successive creations. It fell to the colourful and myriad varieties of tropical birds and beetles to inspire both Charles Darwin and Alfred Russel Wallace into independent but identical proposals of the theory of Evolution by Natural Selection. It became Darwin's lot to amass and organize the enormous volume of information that was needed to erode the traditional views of species immutability and to support the new theory of species evolution.[1]

Darwin's books *The Origin of Species* and *The Descent of Man* remain classic examples of how to present a scientific argument. Drastically condensed, Darwin's theory can be summarized as follows:

1. Living forms show great variation within species.
2. Living forms have enormous numbers of young.
3. Living forms tend to remain at the same numbers, therefore there must be very high mortality of these young.
4. Offspring tend to resemble their parents, so we can infer that the source of much of the variation we see lies in hereditary material present in the sperm and egg of the parents.
5. Those forms of variation that prove to be the most fit or successful will tend to survive at the expense of those that are less fit.
6. The hereditary characters that are responsible for these 'adaptive' characteristics will therefore be selected by nature and be passed on to subsequent generations.
7. Species form can thus change to suit environmental conditions. Where a species evolves to suit different environments more than one form may result and if breeding between these different varieties ceases (genetic isolation) for any reason two different species may evolve.

Darwin's theory proposed that not only should man be classified with the living animals he most resembled, the apes and monkeys, but that he must be viewed as being biologically related to them. Moreover man must have shared a common ancestor with the apes, monkeys and lower primates in the recent geological past and have successively older shared ancestors with other mammals, reptiles and birds, amphibians, fish and even invertebrates. Current opinion might well have accepted a certain amount of adaptive mutability in the basic created forms of life, but Darwin's theory went much further than this. It proposed that man, far from being created in God's image, was simply the end product of countless chancy survivals and mortalities, that through the eons of geological and climatic changes our ancestors had maintained sufficient fitness to throw up a descendant in the biological form of modern man.

Darwin knew nothing about the way that hereditary units were passed on. He was not aware of the work of the geneticist Mendel. Nor were there, in Darwin's time, any known fossil intermediates between men and apes, the missing links that his theory insisted must have existed. Today we know about the laws of heredity and even understand a great deal about the biochemical structure and replication of the DNA molecule by which hereditary information is coded on the chromosomes. But apart from our increased understanding of genetics, we now have hundreds of fossils of prehistoric men and apes with which to back up our arguments. It says an enormous amount for Darwin's thoroughness that despite a hundred years of exposure, discussion and further research, his main thesis remains essentially unchanged and totally acceptable. Superfamily Hominoidea stands unchallenged.

Orthodox classification divides the apes into three families: the Hylobatidae (gibbons or lesser apes), the Pongidae great apes (chimpanzees, gorillas and orang-utans) and the Hominidae (man). The original classification was based entirely upon the anatomical morphology of the various species with particular reference to their spines, limbs, skulls, jaws and teeth. It was originally imagined[2] that this classification reflected evolutionary history and that small brachiating apes of the lesser ape grade gave rise to large, semi-terrestrial great apes, from which there arose a line of terrestrial bipedal apes of a human or hominid grade. This theory can be represented by a simple family tree.

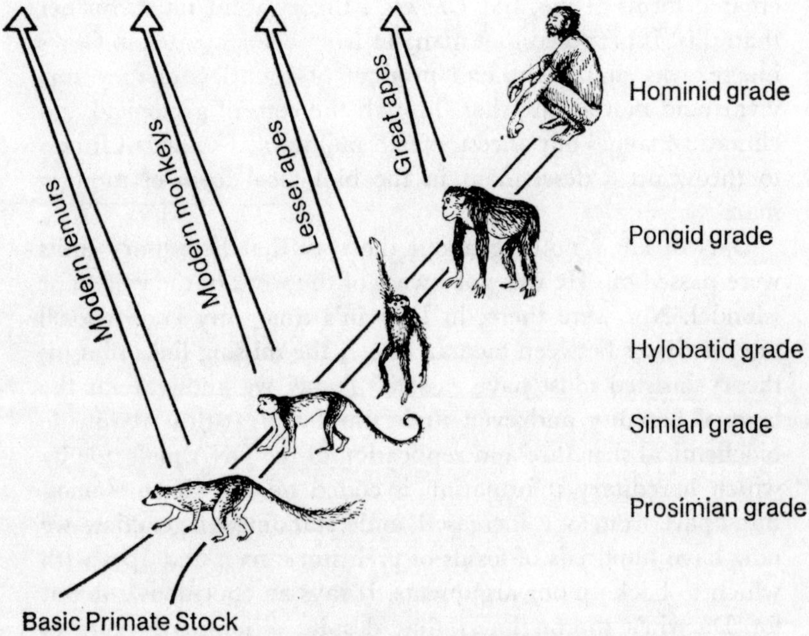

FIG. A. Early hypothesis of Human Evolution

Fossils give a discontinuous picture of evolution. Suitable conditions for (a) preserving fossils and (b) exposing them are rare. We have occasional glimpses into the past but we rarely find a continual process set out. Every link one finds in an evolutionary tree creates two more gaps to fill and so it is with our own order, the primates.

Our oldest fossil primates from North America and Europe, an incredible sixty million years old, are primitive, long-snouted, insectivorous animals, in appearance rather like modern tree shrews, which would command little attention if we did not know with hindsight that they were the ancestral stock from which all the modern primates have sprung. In Oligocene deposits, some thirty million years old, from Fayum in Egypt, we find more clearly recognizable primates in a variety of lemur-like and monkey-like forms.

It is not until the Miocene, some twenty-two to fourteen million years ago, that we find the first recognizable tail-less apes belonging to three families: the oak apes (dryopithecines), tree apes (dendropithecines) and Rama apes (ramapithecines). These families have been proposed as ancestral great apes, lesser apes and men-apes respectively.[3] Slightly later fossils from Asia and Europe also reveal several species of the same three families.

By the beginning of the Pleistocene, a mere three million years ago, there were at least two branches of fully bipedal ape-men. Both lines survived until about a million years ago when they became gradually replaced by a third hominid line of recognizable, large-brained man, allocated to a species succession of *Homo habilis*, *Homo erectus*, and finally modern man *Homo sapiens*. No recent fossils of the African apes have been found but we do have Pleistocene fossils of ancestral orang-utans and giant apes (*Gigantopithecus*) from China and of orang-utans, siamangs and gibbons, together with early man, from Java.

In short, we have numerous intermediates between man and the earliest primates. The fossil spectra show animals of successively progressive organization, indicating that man's ancestry did indeed pass through an ape grade and a monkey grade, though probably never a brachiating gibbon-like grade. What the fossils do not tell us, however, is which precise fossil in one horizon gave rise to which fossil in another. When the Miocene oak ape finds included a giant species and a smaller species it was only too tempting to think of these as proto-gorilla and proto-chimp respectively. With subsequent discoveries the number of oak ape species rose to three, then five and finally eleven, and it became increasingly unconvincing to interpret these as prototypes of today's apes. It is here that biochemical technology can come to our aid.

Biochemists have developed several methods by which we can measure similarities between modern apes and man, enabling us to construct a biochemical family tree which ought to be consistent with our interpretation of the fossil record. Measures of similarity of blood groups, serum precipitation rates, the molecular structure of haemoglobins, myoglobin and other proteins, and comparisons of chromosomal karyotypes and even of the hereditary codings contained in the DNA molecules can be made. All these biochemical methods give similar pictures of the relationships within the living apes, namely that the African apes – gorilla, chimpanzee and man – are remarkably closely related, whilst the orang-utan is quite distinct and the gibbon is even more distantly related.[4,5,6] This contrasts with the orthodox classification which would put the three great apes close together with man rather distinct, and it implies that man separated from the evolutionary line of gorillas and chimpanzees more recently than did the orang-utan.

In theory we should be able to put actual datings to these evolutionary separations since we believe that the structure of many of the proteins has undergone slow evolution at a relatively steady rate. However, the rates at which new variants of functionally identical proteins can become incorporated by whole species must be measured in numbers of generations, not in absolute years. Generation periods of man are longer than those of great apes, which are longer than those of lesser apes; which are in turn longer than those of monkeys, lower primates, etc. The higher up the family tree one gets towards man, the slower becomes the expected rate of biochemical evolution. By taking rates of molecular evolution that have been established for other vertebrate groups and applying these to the apes, making appropriate allowances for these differential generation periods, we can construct a dated genealogical tree that is consistent with the fossil record.

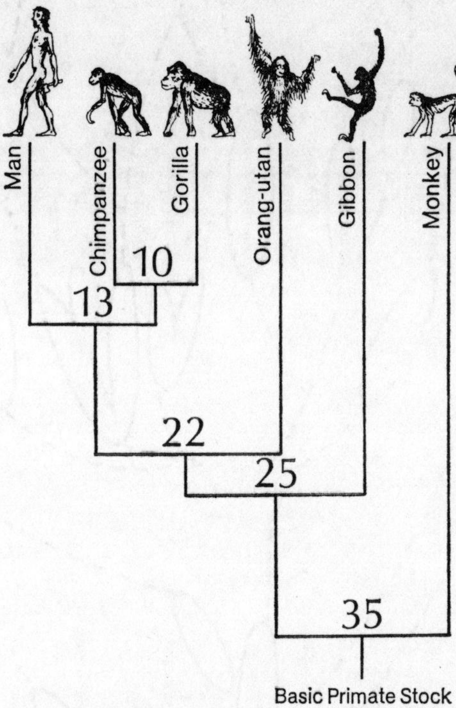

FIG. B. Biochemical family tree with divergences dated in millions of years

This genealogical tree is also consistent with our knowledge of the geological and climatic history of the areas concerned, for throughout the many millions of years over which the apes have evolved, the world itself has been in a constant state of flux. The great land masses were not arranged as they are today; tectonic movements of the earth's crust have caused the

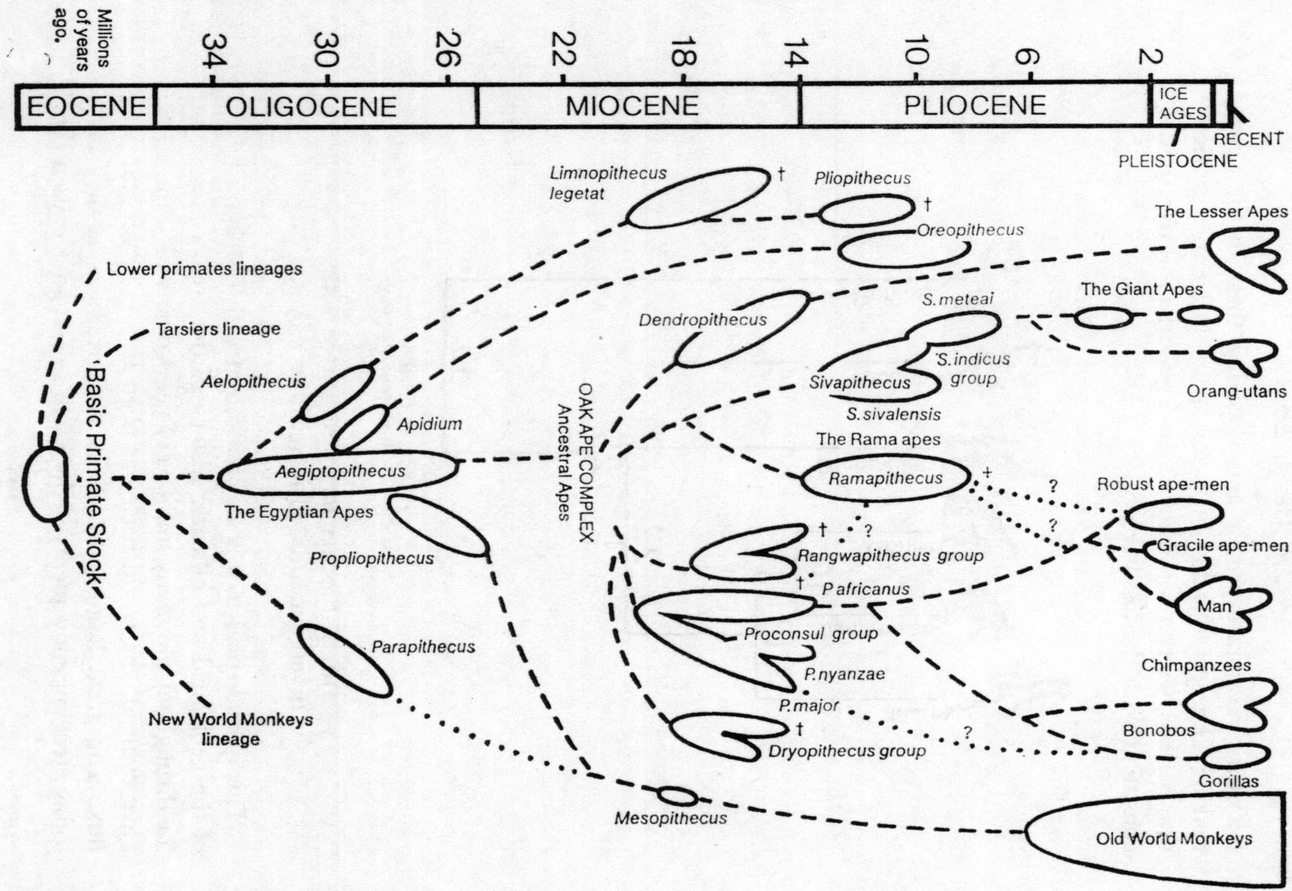
Millions of years ago.
34
30
26
22
18
14
10
6
2
EOCENE
OLIGOCENE
MIOCENE
PLIOCENE
ICE AGES
RECENT
PLEISTOCENE
Basic Primate Stock
Lower primates lineages
Tarsiers lineage
New World Monkeys lineage
Aelopithecus
Apidium
Aegiptopithecus
The Egyptian Apes
Propliopithecus
Parapithecus
OAK APE COMPLEX
Ancestral Apes
Limnopithecus legetat
Pliopithecus
Oreopithecus
Dendropithecus
S. meteai
S. indicus group
Sivapithecus
S. sivalensis
The Rama apes
Ramapithecus
Rangwapithecus group
P africanus
Proconsul group
P. nyanzae
P. major
Dryopithecus group
Mesopithecus
The Lesser Apes
The Giant Apes
Orang-utans
Robust ape-men
Gracile ape-men
Man
Chimpanzees
Bonobos
Gorillas
Old World Monkeys

continents to move together and separate again like floating tea-leaves. Land masses have been torn apart and been flooded by new seas; new land has emerged out of the sea. These geological changes have been of great importance in splitting old evolutionary lineages and stimulating new directions of evolutionary progress.

Only when we bring together our knowledge of the geological events, the fossil record and our biochemical relationships are we at last in a position to reconstruct and interpret the evolutionary history of our ape super-family from a behavioural and ecological viewpoint.[7]

Our story begins in the lush forests that cloaked much of North Africa during the Oligocene, some thirty million years ago. Nascent ape, long-tailed *Aegiptopithecus* (the Egyptian ape), scampered along the ground and clambered among the main branches. His dentition equipped him as an omnivorous, opportunistic, but primarily vegetarian feeder. The Egyptian apes lived in loosely-organized social communities with long-toothed males competing for rank and leadership. If the whole picture sounds rather monkey-like that is not surprising. The whole forest environment was ideal for monkey exploitation as are the tropical forests today, where monkeys are numerically and by mass the most abundant of all mammals. The structure and food distribution of tropical rainforest constitutes an evolutionary womb, inviting monkey evolution. Our ape ancestors were nature's first attempt to fill this monkey niche. Living in the same forests as Egyptian apes were other, smaller, rather lemur-like primates. These creatures form no part in our apely family tree yet they are of tantamount importance to the evolution of the apes and the eventual emergence of man. In their arboreal garden of Eden the Egyptian apes were under little pressure to socialize, they could be as selfish as they wished and remained socially primitive.

Whilst the Egyptian apes enjoyed the easy monkey life in the rainforest with little competition for arboreal foods, some of the smaller primates were biding their time in more terrestrial open-country niches, improving their basic anatomy, evolving more advanced arterial design, skull structure and dentition. Most important of all, however, they were evolving a tighter, more regimented social organization. Just as open-country terrestrial habits encouraged group living in the mongooses, ground squirrels, hunting dogs, and other mammal groups, so it was the more terrestrial, open-country primates that made the biggest advances in social organization.

When, some ten million years later, these advanced social primates, the Old World monkeys, eventually began to compete with the Egyptian apes for the prized forest niches, it was the former, nature's second design of monkeys, that won the day. If they had not, or had perished in the open bush, our own Aegiptopithecine ancestors would have remained forest monkeys for ever. It was their failure at playing monkey that forced them on the hard road to 'apehood'.

In the New World the inevitable patterns of evolution were playing out a parallel story. Continental drift had torn North America away from the super-continent of Eurasia, but not before it was inhabited by the same basic lemuroid primates that roamed over both Asia and Europe. By island-hopping across the shallow seas of the Panamanian region these primates gained early access to the tropical forests of South America. Here again their evolution was moulded in the rainforest womb to fill the monkey niche. The first attempts were the New World monkeys we know today but, as in Africa, there was a second split-off of smaller, though in some ways more advanced, primates which became monkeyfied and competed with the established monkeys. Had the ancestors of the marmosets won this competition, we might have seen New World apes and maybe even New World men, but it never happened. The New World monkeys, with their advanced social organization and locomotor abilities, were already too far ahead in the monkey race to be supplanted. It is obviously hard to displace an animal that can hang on with its tail! The marmosets survived only by filling specialized dwarf monkey niches vacant because of the absence of lorises in America.

Meanwhile, in Africa the descendants of the Egyptian apes were following the classic pattern of redundant biological models. They were forced to specialize within the forest or occupy less desirable habitats elsewhere, in the swamps, mountains and deserts. The garden of Eden days were over and from now on the apes and budding man were going to have a hard time of things.

The long years of competition with the Old World monkeys for the central forest niches left an indelible mark on our Miocene ancestors, for in diverging ecologically from the monkeys they became apes: large, brainy and tail-less.

The Egyptian apes had always been larger than the ancestral Old World monkeys. Size was their main competitive advantage. Although they could not prevent the more efficient social monkeys from out-eating them, they could at least avoid being chased out of the best food trees. As the evolving monkeys got gradually larger so the ancestral apes kept ahead to maintain the differential.

But with increased size arboreal travel in the monkey style became increasingly dangerous and the apes had to modify their locomotion. Like modern monkeys, the proto-apes ran quadrapedally along the top of the branches, leaping from one bough to another, using their long tails to help them keep balance. The problems of such precarious balancing became greater with increasing size. The largest arboreal monkey today, the proboscis monkey of Borneo, shows modifications to typical monkey locomotion that parallel those forced upon the proto-apes. For instance, the proboscis monkey shows a marked tendency to rest and travel in a suspensory rather than quadrupedal manner, with much of its weight borne by the arms holding overhead branches. This is precisely the mode of travel that the apes had to adopt. They left the acrobatics to the smaller, more agile monkeys and became slow, careful, suspensory climbers. The monkey-like tail was no longer needed and became reduced to a stump before it eventually vanished. (For exactly the same reason the tail was lost by the slow climbing lorises and pottos, sloths and even bears.) However, the most important consequence of this change in locomotion was that it demanded a change in body posture and organization. In quadrupedal animals like monkeys the spine is held in a near-to-horizontal position, the guts are suspended beneath this bridge and the muscles of the back must be strong to take the strain. In a suspensory animal, or a vertical clinger and leaper, the spine is more upright and the guts must be supported from the rib cage, with the strain going down rather than across the spine. The ape spine became a pillar instead of a span and the stage became set for a great radiation of apes: bizarre variants on the upright theme – apes to hang from long arms, apes to stand upright on two legs, apes to clamber through trees, apes to hop through swamps, quadrupedal, knuckle-walking apes, giant apes, dwarf apes, fruit-eating apes, meat-eating apes, herb-eating apes, seed-eating apes.

Before we discuss the ape radiation of the Miocene period, however, we should stay a little longer with our original proto-ape, for he is the vital link in understanding our relationship with the living apes. As yet he remains a shady and hypothetical creature; he needs some flesh and hair, he needs a face.

The proto-ape we are trying to reconstruct must share all the common denominators of the surviving men and apes. He must have been sufficiently generalized anatomically and socially to have given rise to the very diverse forms we are left with, and he was probably more similar to the large rainforest, frugivorous apes, the orang-utan and bonobo, than to those types that have departed further from the ancestral rainforest niche.

We have already inferred that anatomically he was larger than a monkey, tail-less and with a tendency to maintain upright rather than horizontal body postures. He must have had powerful gripping hands and feet for climbing through the forest canopy, but could brachiate, walk quadrupedally or even bipedally if needed – modes of locomotion which despite their respective specializations all men and apes can still perform quite effectively.

Socially, we have argued, the apes are essentially both primitive and flexible. The loosely-organized community system seen in Sumatran orang-utans makes the best common model from which to derive all the other ape societies. It is very similar to the primitive social organizations shown by such canopy-living lower primates as lorises and bush-babies. Our proto-apes must have led rather solitary lives: mothers accompanied only by their young but consorting for short periods with somewhat larger males, who may have remained with them during the early infancy of their young; but males otherwise ranged alone.

Since all great apes make nests and man makes beds it is quite likely that this behaviour is commonly derived from our proto-ape ancestor. Nest-making is also common among many lower primates, including the bush-babies and mouse lemurs. The lesser apes probably lost this primitive habit because their smaller size made nesting unnecessary. They sleep like monkeys sitting on large branches.

Our proto-ape probably sported a coat of long coarse hair like that of modern apes and as he faced few arboreal predators we can infer that his coat was black – the easiest pigment to produce and the colour retained by the most primitive surviving members of almost all mammalian groups. Like all modern apes and man, he probably had a naked, expressive face capable of producing meaningful gestures and expressions for short-range communication. At long range he kept in touch by loud rhythmical calls amplified and modulated by an inflatible throat pouch. Such pouches are still seen in the orang-utan and the more primitive species of lesser apes and signs of a vestigial pouch and laryngeal sacs are evident in the African apes and man.

We have no fossils that precisely cover the probable date of our proto-ape but the earliest fossil oak ape *Proconsul* from Kenya, at an age of twenty million years, only slightly post-dates our common ancestor. The anatomy of *Proconsul* closely matches our behavioural and ecological predictions, being larger than a monkey, clearly developing as a branch climber but still very unspecialized in locomotor abilities, with moderate size differences between the sexes, and still rather primitive dentition.

About twenty-five million years ago our proto-ape lived in the continuous forest belt which linked North-East Africa and the southern shores of Eurasia, somewhere in the region of southern Arabia today. But soon afterwards a tremendous rending of the earth's crust began tearing Arabia apart from Africa. The developing Red Sea acted as a partial barrier to faunal movement to the north, while the ever-widening Tethys Sea completely separated Europe from Africa and southern Asia. The climate too was changing, becoming gradually drier, so that the rainforest connection between Africa and India was broken. Some flow of more open woodland and savannah animals continued between the two continents, including some of the advanced Rama apes, but as the Gulf of Aden widened, the corridor became narrower and drier until finally the Indian Ocean broke through to flood the Red Sea some twelve million years ago.

The European oak apes had become extinct as the climate became more temperate but in Asia oak apes survived and at least three Pliocene species can be distinguished. These make good ancestors for the orang-utans and Chinese giant apes, but the ancestry of the Asian gibbons is far less clear. Most authors regard the lesser apes as derived from the African and European tree apes but no fossil ancestors have been found in Asia. Biochemical similarities between gibbons and other apes particularly in the myoglobin molecule argue against such an early split-off.[8] Moreover the evolutionary direction from tree apes to gibbons is behaviourally improbable. We have seen how it was their increased size that forced the change in body posture on the apes. Only if the gibbon was a dwarf ape and had passed through a larger phase would we expect it to show the type of tail-less brachiation that we find in modern gibbons and siamangs. Had the gibbons been evolved from the small tree apes they would never have given up so useful an appendage as a tail. More likely they would have developed it into a prehensile fifth limb like the South American spider monkeys.

There is further evidence to support the idea that gibbons are dwarf apes. Among the living primates longevity and gestation period are closely correlated with body size, the only real exception being the gibbons, who have the life span and gestation periods one would expect of a far larger animal. The gibbons' high intelligence also suggests that they may be dwarf derivations of a larger ancestor. Moreover it is the siamang, largest of the lesser apes, who shows the most primitive features, indicating that the direction of evolution in the hylobatids has been towards smaller size. It seems more realistic to derive the gibbons from the base of the oak ape complex. The absence of the ancestral gibbons from fossil collections is not so surprising since none of the Asian Pliocene fossil beds yet discovered represents fauna of closed rainforest, which is the gibbons' exclusive habitat today.

The Pleistocene era of the last three million years was a turbulent affair of bitter ice ages alternated by mild interglacial (or pluvial) periods. During the ice ages much of our planet's water was held in polar glaciers and sea levels dropped by about two hundred feet. The shallow seas of South-East Asia were drained and the Sunda islands of Java, Sumatra and Borneo all became connected to the Asian mainland. This newly-exposed Sundaland provided a tropical refuge for mainland forest animals including the orang-utan and lesser apes.

Although the area of rainforest shrank during the cool, dry, interpluvial periods it never disappeared and when the last ice age ended the forest spread north again with its diverse flora and fauna still intact.

Back in Africa the Rift formation processes continued throughout the Miocene. The most dramatic rent from the Nile to the Zambezi along the line of Lakes Rudolf, Edward, Kivu, Tanganyika and Rukwa created a formidable barrier to animal movement. This line of steep mountain escarpments, crocodile-infested lakes and swampy valleys completely separated the apes of West Africa from those of the east. Moreover the African climate is such that the line of the Rift almost precisely delineates the distribution of rainforest. To the west the forests extend uninterrupted to the Atlantic, but to the east they are replaced by savannah woodlands. The apes on either side of this barrier inevitably proceeded to evolve in quite different directions as forest apes and bipedal men-apes respectively. We have not a single fossil from West Africa to help us follow the course of evolution taken by the forest apes but three species, gorillas, chimps and bonobos, have survived there. All are terrestrial, quadrupedal knuckle-walkers and so similar biochemically that many authors would like to include them all in the same genus.

In West Africa the effects of the dry Pleistocene glacials were more severe than in South-East Asia. The rainforest almost totally vanished. Together with the rainforest trees died any arboreal apes there may have been, and only the terrestrial knuckle-walking apes proved sufficiently adaptable to survive the changes.

From biochemical data we have to infer that the gorillas and chimps became isolated very soon after the closing of the Great Rift system. Probably the barrier between them was the Niger-Chad river system, which was formerly a major waterway and a barrier to many other animals. The chimp occupied the knuckle-walking ape niche on the west of the Niger and the gorilla replaced it on the east. The two species probably evolved along very similar lines until that fateful day when the river barrier dried up during the Pleistocene interpluvial, permitting the chimps to cross back into the gorilla's world. The smaller ape, the chimpanzee, proved to be ecologically superior and the gorilla was forced to diverge in its habitat use or face extinction. The gorilla survived the re-exposure to chimps by becoming a terrestrial giant, exploiting the lush herb layers of secondary forest patches, leaving the canopy strata for the more agile chimps.

Although it eats very little fruit in the wild today, the gorilla obviously had a recent fruit-eating ancestor. In feeding experiments captive gorillas consistently show preference for fruits over other types of food. The gorilla's broad incisors and simple, though long, intestines are evidence of a fruit-eating past. His fibrous dung indicates his inefficiency at digesting the tough plants he now has to eat and to obtain enough nourishment he has to consume enormous quantities of these herbs. Although the gorilla diverged from the chimp sufficiently long ago to increase his size and modify his behaviour, he has not had time to lose his sweet tooth or evolve a complex, browser's gut.

The recent nature of the gorilla's gigantism is also revealed by the fact that his longevity, period of gestation, surface area to weight ratio and brain size are all characteristic of a smaller animal than modern gorilla. The very loosely-bonded nature of gorilla groups also suggests recent origin of group-living in this species. Groups are held together by the individual pair bonds of different females to the central dominant silverback. Remove the king-pin and the whole gorilla group collapses. The gorilla has not been group-living long enough to evolve the tight network of social bonds that make monkey groups so stable.

Lava dams across the Great African Rift system have only recently permitted African forest apes to cross into East Africa again, for the first time since the Miocene, but apes have not advanced far. Gorillas traversed the Virunga dam into Rwanda and Uganda. Chimpanzees crossed the Ruwenzori dam into Uganda and the Bukavu dam to spread down the eastern shore of Lake Tanganyika. Chimps worked right round the headwaters of the Lualaba river to colonize the Congo Basin but the bonobo had been isolated long enough to have evolved as a separate species, as a specifically rainforest inhabitant. On the eastern side of the African Rift system ape survival was a far more dodgy affair. Oak apes became extinct during the drier conditions of the Pliocene and only the men-apes survived through to the Pleistocene. The almost complete lack of fossil-bearing deposits of between eleven and three million years' age is unfortunate, for it obscures the exact origin of the men-apes. The canine reduction evident in the first jaw fragments found of the Rama ape made him hot favourite as an ape-man ancestor. More complete fossils have revealed that he did not, as was first thought, have a rounded jaw and it now seems that his origins lie among the Asian apes rather than in Africa. If the Rama apes still turn out to be human ancestors we must accept a few kinks in our evolutionary course but perhaps we have not yet discovered our Pliocene ancestor. Biochemical similarities between man and the African apes indicate a recent divergence, but in the absence of any stronger contender we should look to *Proconsul*, the oldest and least specialized of the oak apes, as our most likely common ancestor.

After all the efforts of palaeontologists and archaeologists, we are left with a gap in man's origin, a missing link almost as wide as that which so frustrated the Victorian Darwinists, a curtain of eight million years over that vital transition from a quadrupedal ape to a fully bipedal primitive ape-man. No wonder scientists have tried so hard to squeeze the Rama apes into the void. As it is, we can only theorize about the evolutionary pressures that brought about the transformation from ape to man.

The climate of South and East Africa became progressively

drier throughout the Pliocene. Much of the area was wooded savannah and open deciduous woodland, but permanent greenery was found only in the gallery forests along waterways and in the marshy woodlands surrounding the shallow Rift lakes. True grassy savannahs were probably not common but are recent artificial phenomena that have resulted from the annual fire-spreading activities of pastoral man to stimulate new grass growth.

What happened to those forest apes that were trapped on the wrong side of the Rift barrier when the forests began to thin? They might well have become extinct but we know they did not, for when the eight-million-year curtain was lifted their descendants are revealed as a lineage of fully bipedal large Robust ape-men (*Paranthropus*) and a second lineage of smaller, more Gracile ape-men (*Australopithecus*). The origin of the bipedal hominids has often been conveniently evaded by vague references to open-country living – a transition from fruit-eating to hunting.[9] Bipedalism has been explained as an economic way to travel long distances with the added advantage that it leaves the hands free for carrying, and using tools and hunting weapons. The shortening and rounding of the jaws seen in the fossil series were said to reflect a change in diet from a herbivore, with his tools in his jaws, to a carnivore with his tools in his hands. Actual changes in hand structure have been interpreted as permitting increased use of a precision grip for holding objects between the thumb and forefinger rather than the ape power grip for holding objects in the palm of the hand. The development of high intelligence and speech were explained as vital for greater efficiency in co-operative hunting techniques. The argument is circular, it all fits together, and the various adaptations are all mutually encouraging, but however did a vegetarian quadrupedal ape ever get itself into such a spiral? The hunting chimpanzee analogy does not really help. The chimp uses his speed and acceleration to capture prey; he is fastest on four legs, not two, and can run far faster in rough terrain than a modern athlete. He uses the strength of his hands to kill his prey, hitting the victim or picking it up and smashing it on the ground. He does not use tools for killing or disembowelling, indeed he could not open up large prey at all were he not equipped with his sharp long canine teeth. Female and young chimpanzees are unable to open up baboon corpses and must beg for meat from the dominant males. If the chimpanzee were forced into a greater dependence on hunting, we would expect this ape to develop even faster quadrupedal locomotion and longer, sharper canines. For an ape to start killing and disembowelling prey with stone tools rather than his teeth he would need to have already developed a tool-using culture far in advance of anything seen among modern chimpanzees. That such a culture already existed in our Miocene ape ancestors is unbelievable.

From the ape-men fossils we know that bipedalism and the reduced dental arcade preceded the advent of stone tools and that marked increase in brain size came even later in our evolutionary history. One could argue that the tools being carried by the early bipeds were either of perishable wood or so crude as to be unrecognizable, but if this were the case then they were not used for hunting. Only a long, sharp piece of cracked stone or bone would be suitable for opening up the belly of a large animal but no such remains have been found. It seems most unlikely that it was a change to open-country hunting that led to the development of bipedalism or the reduction of front teeth.

There is obviously a piece missing in the jigsaw and the most convincing proposal so far suggests that the early hominids became adapted to a staple diet of grass seeds and root tubers.[10] Their ape ancestors already preferred eating fruit but as the woodlands dried up fruit was in short supply. Moreover since fresh green shoots were available only seasonally to satisfy their water requirements, our proto-hominids were obliged to live around seasonal and permanent water-holes. In a wooded landscape this is precisely where one finds the greatest abundance of grasses. Grass seeds provided a rich food supply for which there was little competition. But individual grass seeds are so tiny that it is difficult for most large animals to collect them in sufficient quantity to constitute a substantial part of their diet. For a fruit-picking ape, however, the task was quite easy and the gradual evolution of a more effective precision grip, with the thumb opposite the first finger, further facilitated the delicate operation of handling such small food items.

Changes in dentition shown by the fossil apes are also consistent with the job of crushing and grinding small hard seeds. The front teeth and canines became reduced so that rotary chewing was easier, whilst the back teeth became flattened to crush and grind like a miniature mill wheel. The loss of the canines as weapons could be tolerated by an ape which waved and threw branches or stones, or wielded large sticks as clubs, an ape that had enjoyed a tool culture well within the limits of modern chimpanzees.

Collecting enough tiny seeds for subsistence must have taken our early hominid much of the day but he would not need to move far. It would therefore be his feeding posture rather than locomotor patterns which determined how his anatomy developed. Grass grows tall and we can imagine our grass-collecting apes stretching up to reach the seed heads. To free both hands for collecting seeds, a quadrupedal ape must sit on its haunches, changing his position every few minutes to reach new grass heads. This meant standing up on all fours and taking a few paces before settling again or else shuffling clumsily about on its bottom. It would also be wise for the feeding quadrupeds to stand up bipedally every few minutes to look out for approaching predators just as chimpanzees do today. The premium for long periods of bipedal standing suddenly becomes obvious. The standing apes had both hands free to pick seeds and their heads clear to keep a permanent look-out. It did not matter that bipedal locomotion was not at first very efficient, the animals only needed to move a few yards every few minutes while feeding. If speed was necessary our apes could still drop back on to all fours and scamper away. We can see exactly the same process at work in our own children. It wasn't until he wanted to reach some blackberries that my infant son David, until then a confirmed crawler, decided to become bipedal.

As bipedalism became more efficient, our seed-eating apes may have taken to carrying wooden sticks with them, sticks which would double up as weapons and useful digging tools for scraping out termite nests, edible rhizomes and tubers. The Gombe chimpanzees will carry grass termite tools for several hours before actually using them and even a quadrupedal knuckle-walking ape can carry a tool in its hand while walking.

Such a seed- and tuber-eating way of life, although possible for a quadrupedal ape, would undoubtedly favour the development of a crude tool culture, bipedalism, precision-gripping hands and hominid dentition without greatly straining the ape intelligence. The Robust ape-men can be seen as the direct end-point of such an evolutionary line but the intermediate vegetarian hominid stage was beautifully pre-adaptive for the cultures of more advanced tool industries, and the eventual hunting techniques embarked upon by the Gracile ape-men and later branches of *Homo* that finally led to man. Meat-eating obviously played an increasingly important part in the way of life of the Gracile ape-men and later true men, but with the women-folk still collecting large quantities of seeds, tubers and vitamin-rich fruits to supplement the hunter's bag. Our ancestors had reached the hunter-gatherer way of life but as with surviving hunters today, Eskimoes, Pygmies, Bushmen, Red Indians, the bulk of the diet was probably carbohydrate rather than protein. To this day grass seeds (rice, maize, wheat, oats and barley) and root tubers (potatoes, yams and manioc) still form the staple diet of almost all human races.

Although hunting may not have been the behavioural

change that brought about bipedalism, it did play an enormous part in the shaping of our species' social behaviour, early culture and even anatomy. Our striding gait, and extraordinary arm flexibility are clearly the inheritance of the hunter. Man cannot run as fast as the fleet-footed plains herds, nor is he outstanding at jumping or swimming, but he can walk a very long way in a day and he can hurl a spear from an accelerating run with unbelievable power and accuracy. No ape can throw a spear, but man can not only deliver the shaft in a straight line along its axis but on release his fingers automatically impart a rotary spin which is essential to stabilize the spear's flight through the air. To throw a spear well a man needs all-round good physical development, no part overdeveloped but every part supremely fast and athletic. The perfectly proportioned Adonis shape is the ideal physique for the modern Olympic javelin thrower and we must conclude that of all man's athletic abilities the function for which he is most perfectly designed is throwing a spear from an accelerating run. His ability to throw a lethal 'fang' at one hundred miles an hour over distances up to a hundred yards made man the most efficient killer of large game since the giant carnivorous dinosaur Tyrannosaurus became extinct seventy million years earlier.[11]

As our ancestors were forced to live in increasingly arid environments they lost most of their body hair. Many authors have made a great deal out of our species' comparative nudity and suggested ingenious theories to account for our nakedness.[12,13] But we did not lose our hair to make us sexier, or because we were immersed for eight million years in the warm seas of the East African coast. The truth is much less spectacular. We never did have much real hair in the first place and what we did have was quite unsuitable for our new way of life. Hair loss is not a human characteristic but a feature of apes. In most mammals one of the main functions of hair is to keep the body warm by regulating the loss of body heat. When the mammal is cold it erects its hair slightly, thus increasing the layer of insulating air that is trapped by the fine soft under-fur close to the animal's skin. This reduces the rate at which body heat is lost. If the animal is hot, the hair lies down tight so that little air is trapped and maximum heat is lost from the body. Larger, stiffer hairs called guard hairs protect the soft fluffy hairs of the undercoat and act as tiny rudders to steer wind and rainwater off the body and to protect the animal from sharp thorns. Monkeys have both kinds of hair but not the great apes. These large-bodied animals of the equatorial rainforests have lost the need of a fine undercoat and have modified their guard hairs into coarse, brittle strands that protect the skin from grazing and deflect rainwater off the body.

Look at a chimp or, better still, an orang-utan next time you are in a zoo and you will see how naked it really is. With a hair density of less than two hundred hairs per square inch an ape can form no insulating layer, wind can pass over the skin unchecked and the body hair provides virtually no warmth. When apes are hot they must sweat to cool down and when they are cold they must shiver to keep warm or curl up in a deep leafy nest. They live in the most stable terrestrial environment on earth; the tropical rainforest temperature rarely drops below 22°C (72°F) and rarely exceeds 35°C (95°F). Provided large forest apes keep dry they are not faced with any great problems from cold, they do not need soft under-fur and indeed this would greatly impair their activity. All the large tropical forest mammals that are active by day – elephants, wild pigs, rhinoceros – have similarly lost their under-fur, whilst those that have not done so remain mostly nocturnal in their behaviour – tapirs, tigers, sambur deer and wild cattle. The heat regulation problems of small animals, squirrels, tree shrews, monkeys, etc., are quite different; their surface area is much greater in relation to their weight and consequently they lose heat much more rapidly. They need hair to keep warm at night, yet can still be active during the day without over-heating.

What we now have to imagine is what happened to our sparse, coarse ape coat when the forests of East Africa dried up and our ancestors became the hunter-gatherers of the savannah woodlands. The savannah shows much greater temperature fluctuations than the rainforest. Night temperatures drop to only 16°C (60°F) whilst midday temperatures may soar up to 43°C (110°F). Small mammals still need their fur to keep warm at night but over-heating in the intense heat of midday has forced them to become nocturnal or crepuscular in their habits. Large animals with thick coats would also over-heat in the daytime but they do not need such hair to keep warm at night. Their larger bulk and fat reserves are sufficient. Large plains animals have thus been able to lose most of their hair and by further increasing heat loss by panting and sweating are able to remain active for much of the day. However, unlike the forest animals who forsook their under-fur, the plains game have done the opposite. They have discarded their guard hairs and retained a very short glossy undercoat, not to keep warm but merely as a protection against direct radiation of the sun. Our ape ancestors had learned to sweat all right, but they had forsaken their undercoat. We could not imitate the glossy coat of the antelopes, lions and horses. We had a problem. Too much sunlight is dangerous; not only does the absorption of the radiation lead to over-heating, the ultra-violet content of sunlight can lead to skin cancer. Our solution to the problem was to protect the skin from ultra-violet light with black melanin pigment even though this increased our absorption of sun heat. The most sensitive part of the body, the brains, had to be protected from overheating by a mane of thick brittle hair, but we simply could not afford to go hairy all over, for that would prevent our sweat system from operating and would greatly impair our locomotion. There is nothing new in this use of protective melanin, even forest apes protect the naked parts of their bodies, faces, hands in just this way.

To complete the answer of what happened to our ancestors'

sparse body hair, we must again consider the functions of ape hair in the forest. First, it acts as a drainage aid. With our new upright body posture the direction of water run-off had changed from the hunched rain posture of the ape, and the direction of hair growth changed accordingly. In the ape hair flows back from the wrists to the elbows which are bent in the rain posture and form water streamlets. In man the hair flows the other way, taking water straight down the arms and off the finger tips. But the importance of the drainage function was greatly reduced. Our ancestors' streamlined upright shape was itself excellent for fast drainage and the breeze of the open savannah enabled the skin to dry much quicker than in the humid forests. The drainage function of hair was thus greatly reduced.

As a protection against abrasive surfaces and thorns the hair's function was also reduced. Walking upright over open country involved far less scratching from abrasive vegetation and the scratches themselves were far less likely to become septic in the healthier drier atmosphere. On top of this came the cultural adornments, thongs, bags and necklaces that proto-man was draping about his body. These would rub on the hairs causing sores and rubbing dirt into the follicles. Anyone who keeps horses will know of the problems of straps and harnesses rubbing over hairy sweaty skin. As the use of bodily adornments increased, so the inconvenience and positive handicap of body hair must have been felt – body hair that no longer had any function in heat control and little value for physical protection or as a drainage aid. Body hair became more of a liability than an advantage and was further reduced. The process never was completed, we still have vestigial hair and I know some people who would make an orang-utan look like the more naked ape.

Present-day racial differences in human hairiness give us further clues as to the adaptive relationships between hair and environment. The hairiest peoples, Veddahs and Indo-

Europeans, were originally forest peoples enjoying the original ape habitat of equable climate where the advantages of hair as a protection against abrasive thorns and branches still existed. The peoples with the smoothest skins of all are those who originally lived in either the hottest or coldest climates, Negroids, Australian aborigines and Mongoloids. The heat and sweat of savannah life must have put very strong selective pressure for hair loss on the early Negroids while the Mongoloids, in their harsh northern climates, wear the most rubbing clothes. The Mongoloids crossed the Bering Strait and spread south into the tropics of both the Old and New Worlds so that we now have Mongoloid peoples that are secondarily re-adapted to tropical rainforest, but although their physiognomy and skin colour have been able to re-adapt to the hot humid forests, their relative hairlessness reveals their northern origins.

The two different lineages of bipedal hominids that evolved in East Africa were adapted to quite different life-styles. The Robust ape-man (*Paranthropus*) has puzzled palaeontologists ever since he was first discovered, but today we can see him as a clearly separate, somewhat parallel, branch of the lineages leading to man. He was a crude, heavily-built ape-man, fully bipedal but less efficient than the striding Gracile ape-man (*Australopithecus*). His teeth were massive and flattened, obviously designed for grinding and crushing coarse vegetation and quite different from the delicate omnivorous teeth of Gracile ape-man. The latter was found with crude stone tools, bone daggers and animal food remains. He was obviously a successful hunter. The Robust ape-man is not found in close association with stone tools, yet he was not only contemporary with Gracile ape-man, he occupied many of the same sites. How could two such closely-related species survive together, why didn't one oust the other? Not only did the Robust ape-man survive three million years of living side by side with the more advanced Gracile ape-man, and later the earliest true men in the form of *Homo habilis*, but he apparently thrived. His remains are more numerous than those of his more advanced cousins and he occurred over vast areas of East and South Africa. The Robust ape-man must be regarded as one of nature's more successful creatures. He must have made very good sense ecologically. The picture of a primitive vegetarian who could only waddle and didn't have stone tools sounds a most unlikely success story and needs much deeper examination.[14]

Our first clue lies in his canine teeth which were small and insignificant. Canine teeth in monkeys and apes are used for aggressive social fighting. In species where both males and females fight, such as the gibbon, both sexes have long sharp canines. In gorillas and orang-utans, where it is mainly the males who fight in social and sexual competition, then we find very marked sexual dimorphism in overall body size and in the size of the canine teeth. The skulls of Robust ape-men show great sexual dimorphism, with the large males having high sagittal crests like male gorillas. But our Robust ape-man is unique in showing such sex size difference yet lacking aggressive canines. The large size of the male Robust ape-men tells us that they must have been socially competitive and aggressive, but if they did not carry their social weapons in their mouths they must have carried them in their hands, they must have had fighting tools. We can see a good parallel to this in the evolution of deer. The primitive deer, including modern chevrotains, have long aggressive canine teeth. In more advanced species the canines were lost only after alternative social weapons had been evolved in the form of antlers. The Asian muntjacs represent the overlap stage and have both long canines and small antlers.

We have inferred that the male Robust ape-man fought his brothers with tools, but as we find no stone tools with his bones we must assume that these were perishable wooden clubs. This would explain why the Robust ape-men skulls were so massively solid and why the more advanced, but more delicately built,

Gracile ape-men could never oust the Robust species aggressively. Try attacking a man twice your own weight who is wielding a large wooden club when you have only a stone hand-axe or a bone dagger.

We can also dispel some of the myths about the Robust ape-man's inefficient bipedalism. Anyone who has seen the bipedal effectiveness of orang-utans and chimpanzees cannot doubt the bipedal capacities of anything with so human a pelvis and femur as the Robust ape-man. To imagine that the Robust ape-man was incapable of the human striding gait is sheer nonsense; even chimps can perform this movement. The Robust ape-man may not have moved as gracefully as the Gracile ape-man and would probably tire more quickly over long distances, but his long flat femur heads point to heavy muscular thighs and we should think of him as a tremendous athlete, accomplished at the short dash and leaping.

Their differences in locomotor adaptations, culture and diet indicate that Robust ape-men and Gracile ape-men occupied slightly different habitats. The Robust ape-man was a woodland dweller, living in the gallery forests along the riverside. He must have eaten ferns, herbs, grassheads, acacia beans, bamboo shoots and probably dug up termite mounds or caught water-snails, raided birds' nests and clubbed to death the occasional young or slow animal.

The Gracile ape-men preferred the savannah and flood plains of lakes. Like the Robust ape-men they could not go far from water, and in the dry season they must have been reduced to ambushing large game at isolated water-holes. They ate fewer herbs and seeds than their Robust relatives but more meat and fruit and had to travel large distances in search of both. The Gracile ape-men still retained a sharp, slightly protrusive, canine tooth. Possibly they needed this for tearing flesh from his prey but it may also have had an aggressive function. We have argued that Robust ape-men settled their differences with clubs. Although Gracile ape-men exhibit less sexual dimorphism and probably showed greater social tolerance, it is inconceivable that there was no status friction. How then did Gracile ape-man settle his disputes? It is most unlikely that he jeopardized life and limb and the whole security of his social group by battling it out with his lethal stone hand-axes and bone daggers. Almost certainly weapons were taboo and the males had to revert to the old chimpanzee weapons of pummelling fists. Biting with the teeth became semi-taboo, the signs of a coward or loser, but necessary as a last defence to prevent the stronger fist-fighter from inflicting too much punishment upon the vanquished. The taboo is still clearly visible today in human behaviour and so is the use of teeth and sharp fingernails as a discouragement to the over-aggressive. The tooth taboo must be very old, for it is clearly visible in the Gombe chimpanzees. In disputes over food, females or pure status, the Gombe chimpanzees pummel one another furiously with their fists and feet, they pull hair and drag their opponents savagely over rough ground, but they rarely inflict wounds with their teeth. They fight, displaying, but not using, their more fearsome weapons. Only in inter-community (proto-warfare) aggression do we see the chimp teeth really used to effect and the savage, potentially lethal wounds that can result.

Gracile ape-men must also have lived in large chimp-like communities in which several males were breeding but depended upon one another for communal hunting and range defence. In contrast, the Robust ape-man lived in smaller social groupings in which one male would largely monopolize breeding. In his last million years of existence, the Robust ape-man showed a curious convergence towards his more man-like cousins, becoming lighter in build and smaller-toothed. The convergence was probably due to increasingly arid conditions when the riverside forests shrank and the Robust ape-men were forced into more open habitats. This must have greatly increased the competition between our two hominids, but by this time Gracile ape-men had begat true man and at some sites remains of all three lineages overlap in time. Man was no longer armed with a hand-axe but with a spear and a bolas. The extermination of the Robust ape-man was now inevitable, but he did not die without leaving his mark on mankind. He has given nightmares to two thousand generations of human children, and the folklore of stupid, hairy, club-wielding forest giants being outwitted by small but canny man remains splendidly vivid all over the globe. The image of prehistoric man courting with a wooden club probably gave as much amusement to the Gracile ape-men and earliest true men as it gives to cartoon artists today.

When, two million years ago, the Suez Isthmus emerged out of the sea to relink North Africa via the South Arabian woodlands to tropical India, man could no longer be contained within the African continent and much of the emphasis of human evolution shifted to the setting of Eurasia. Early man spread east to South China and across Sundaland to the plains of Java. He found two surprises waiting for him in Asia, the Chinese giant apes (*Gigantopithecus*) and a mainland race of huge orang-utans. Fossil jaws and teeth of both these monsters have been found. Without further skeletal remains it is not easy to say quite what sort of animal the giant ape was. From his very large size we know he must have been terrestrial but we cannot tell whether he was a biped or a quadruped. His massive jaws and broad grinding teeth are remarkably reminiscent of the African Robust ape-men and we must assume that the giant apes must have had a similar diet. Possibly the giant apes were also bipeds and, if so, they may have been as much as eight feet tall and an uncomfortable reminder of the Robust ape-man mythology. The extermination of the giant apes must have been high on the list of priorities for the first men to reach Asia and even the large mainland orang-utans did not survive this spree of ape-hunting, so that the orang-utan today is limited to the uninhabited jungles of North Sumatra and Borneo.

Like the Cain and Abel story of Genesis, one branch of proto-man murdered the other and has been repeating the performance ever since. From the day that culture won over brawn man's body has remained evolutionarily static. Different races have shown varied pigments and physiognomies to suit different climatic regions of the globe, but anatomically we have made no further advance. All our evolutionary efforts have gone into brain power. Not to hold our own against nature and the elements – we had been doing that for millions of years with an ape-sized brain – no, we needed our guile and our technology to defend ourselves against the threat of neighbouring mankind, and the best means of defence has always been lethal offence.

Part Three

MEN

CHAPTER 8

Technology and Culture

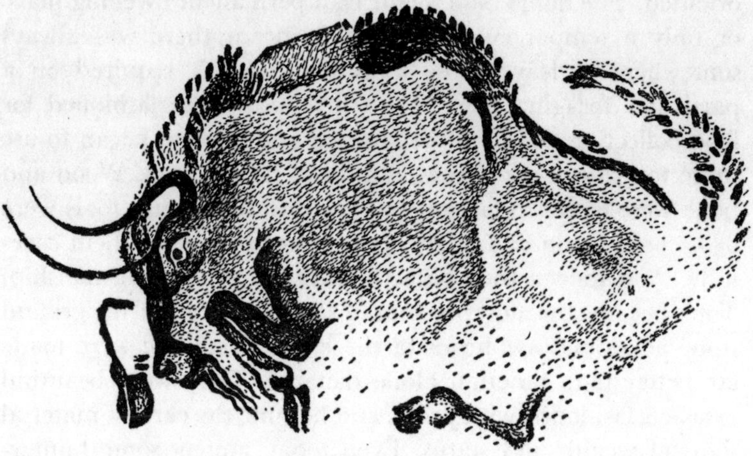

For over ninety per cent of his history man has lived as a hunter-gatherer.[1] From the start his reliance on a tool technology was much greater than anything we have seen in the apes. Apes may fashion a crude tool to solve a particular problem, to use as a lever or hammer, to reach a prize beyond arm's length, to hit a frightening object, to poke or probe into a small hole. They may even foresee that they will require a tool later in the day, prepare it beforehand and carry it around with them; but typically ape tools are used only once and then discarded. Apes need tools so rarely that it is not worth their while to encumber themselves with the additional weight. Moreover since their ranging patterns are so irregular there is no advantage in their leaving tools in places of safe-keeping until they may be required again. The position was rather different for primitive hunter-gatherers, however; they were already fully bipedal. Without any loss of locomotor efficiency, they could easily carry tools in their hands and their dependence on fresh water each day meant that they were base-oriented. The home base might be a permanent dwelling place or only a temporary camp, but it meant there was always somewhere to leave those tools that weren't required on a particular foraging trip. Man's first tools were fashioned for food collecting and food preparation and he soon began to use stone to achieve the shapes and forms he required. Wood and bone were easier to fashion but less durable. Man's tools were to be used again and again. It was worth making them carefully. And indeed they became works of great craftsmanship. Tools became the most precious of all possessions and the ground stone adzes and axe-heads of the late Palaeolithic were made far better than function alone dictated. They were beautiful artefacts fashioned with pride, and became the earliest material signs of wealth and status. Even today among some hunter-gatherers, such as the Hadza tribesmen of Tanzania, tools are still used as a currency.[2] Hadza menfolk use poisoned arrows as the stakes in lengthy gambling sprees even to the point where they have squandered all away and can no longer hunt game.

One of the first inventions of early man must have been the rope, a pliable twine of plant or animal fibre, that could be used to bind and tie objects together so that they could be carried around the neck and waist or suspended from any natural hook. The uses of the rope were so multiple that it became the cornerstone of many future inventions. In less-developed countries twine is still highly valued and even in affluent industrialized societies there survive strong taboos against throwing away string: 'So useful, you never know when you may need it.' We hoard our little balls of twine in equally treasured tins only to discover them untouched years later. Finds in Palaeolithic deposits tell us that Stone Age man already knew about string and used it to thread small stones and hollow bones together as necklaces and to fasten heavier rocks as bolas weights. Probably, however, rope is much older than even the cave-men, indeed it can almost be classed as one of the ape tools. A few years ago there was a switch in zoo practice from providing animals with hay or straw bedding to a more hygienic and less seasonal alternative, fibrous wood shavings. Almost immediately in several different zoos orang-utans began a curious pastime of making rope from the new bedding. One young female I watched at London Zoo was especially proficient at the art. Her game was to carry a bundle of bedding up to the roof, twist it into a long sausage, loop this over one of the bars, twist both loose ends together and then launch off like a pendulum on her rope swing. From string came man's first clothes, a belt or line round his middle from which he could suspend tools. Next came the net bag as a more effective way of carrying not only tools, but also the spoils of the hunt or the roots and berries that were to form the real bulk of his diet. Miamin tribesmen in New Guinea and some African pygmies still wear similar waist thongs made of vines, and net bags still form the only clothing of the Baliem hill tribes of New Guinea. The use of rope as both a functional and decorative body adornment foreshadowed the development of clothing in colder climates. We have seen how irreversibly naked man had become on the African Plains. He could not simply grow a thicker coat of hair when he wanted to move out of the tropics into colder northern climates so he had to take his fur coat second-hand from the large game he hunted.

If any single moment ever enabled a species to increase its impact and dominate its environment it was the day when early man first discovered how to make fire. Given man's ape heritage of intelligence, adaptability and opportunism, the discovery of fire was inevitable. Fires occur naturally during the dry season in savannah woodlands, and must have been a familiar sight to early man. Frequently he must have benefited from the toasted crickets and singed vertebrates left behind the wind-blown walls of fire, just as hawks and jackals do today. He was an intelligent creature who quickly learned the properties of fire, how close you can get without getting hurt, and which side to stand to avoid being trapped or smoked, even that he could drag off burning branches to spread the fire. To a grass-seed eater, fire-spreading became a boon. When scrub is burned grass is encouraged at the expense of other vegetation and by forming a fire mosaic the seeding times of different areas can be staggered throughout the year.

The Gombe chimpanzees know all about scrub fires which sweep over large parts of the reserve every dry season. I vividly recall my own apprehension when a particularly fierce wall of fire was sweeping down 'Sleeping Buffalo' Hill towards our camp. I was amazed to see three old male chimps lumbering down towards the fire and even more surprised when a few minutes later I saw the same animals up on the charred slope of the hill behind the advancing flames. They had hopped through a gap in the wall of fire and once behind the flames were perfectly safe, as the fire travelled too fast to heat the soil or burn the trees. This chimpanzee demonstration showed me how easy it would have been for an intelligent technologist like early man to control fire for his own uses. There were probably countless generations gaining benefit from natural fires before anyone discovered how to kindle flames, but in a culture where man was striking flint against stone to fashion tools sooner or later a spark or glowing splinter would land among a pile of dry grassy bedding and start a blaze.

After a few such accidents, someone must have connected cause and effect and discovered that he could start a fire by directing a barrage of stone chipped sparks into a pile of dry tinder. Later, when the idea of making your own fire was commonplace, it was but a short step to discover that by rubbing two bits of wood together one could kindle a blaze. These two methods of fire-making are still employed in many parts of the modern world where supplies of matches and cigarette lighters are unreliable. Even in the humid rain-forests, where everything in sight looks wet, an experienced jungle traveller can soon start a fire with his bamboo saw and firewood scrapings from one of the aromatic resinous trees. A flame is kindled, blocks of hard inflammable resin are added and soon the fire is hot enough to burn wet wood. Fire broadened the food range of man, gave him an invincible weapon of both offence and defence against the large carnivores and helped him drive the large herds of antelope into traps, over cliffs or into lakes where they could be butchered with relative ease. Man was able to supplant the sabre-toothed cats as the dominant carnivore preying on elephants and other large game. Most important of all it was fire that eventually opened the way from the Stone Age into the metal technologies of the Iron and Bronze Ages.

The discovery of fire is given the highest reverence in the mythology of almost all peoples, ancient and modern, and sun worship was one of the driving forces in the evolution of religion in both the Old and New Worlds. The love of fire is deep-rooted in our species. Whereas most young animals are shy and afraid at their first experience with flames, human babies' eyes light up with glee, they smile, they love the dancing pretty light and they approach dangerously close to play with the delightful new toy.

With fire as his companion man could range the globe. By then the corridor between Africa and Asia was open again for faunal exchange. Man had already spread right across southern Asia to Java but now he had fire he was no longer shackled by his tropical bonds and could spread north into stimulating new environments where bears and wolves ruled the temperate forests. Half a million years ago Peking Man was undoubtedly using fire, and since that time human dwelling sites can be recognized from charred food remains. Fire was useful for cooking vegetable and animal foods as well as for warming human bodies and soon became symbolic of the home, the hearth and man's supreme uniqueness.

Had human history been confined to the limits of Africa it is unlikely that we would have advanced culturally beyond the Stone Age hunter-gatherer grade. The challenge of new environments forced man to innovate, to make new local discoveries, social and cultural advances that spread to other groups. In the tropics man had been able to survive with the aid of a few tools, tools that merely amplified the actions of his own hands, but to endure the harsh northern winters he was forced to adopt such refinements as houses, clothes, fire and storage containers and develop the concepts of planning for the future and rationing winter stores. The climatic instability of the Pleistocene kept our adaptability up to scratch and gave us no chance as a species to stagnate in conservatism. While other animals ranged north and south following the moving climatic belts man could stand still and modify his ways. In the same caves we find human hunting remains in one strata of mammoths, woolly rhinos and bears, and in the next strata of horses and tropical elephants, evidence that, while all around him climate and fauna changed drastically, man could adapt to suit his circumstances. We had already become the most widely distributed land mammal on the planet. But we didn't stop there.

Cultural development depends heavily on communication. The individual learns from other individuals and in turn passes his knowledge on to yet more. As apes, however, our ancestors already had a very well developed ability for communication, and from this human language probably evolved. Almost everyone who has worked with captive chimpanzees has recognized the very high level of understanding these animals show for human gestures, tones and verbal language. They can be readily trained to carry out simple vocal commands, yet they are unable to talk back. The main authority on great apes at the beginning of this century was Robert Yerkes who concluded[3] that: 'If the imitation tendency of the parrot could be coupled with the intelligence of the chimpanzee, the latter undoubtedly could speak.' But although the chimp shows tremendous imitative capacity for action, he is unable to perform the intricacies of human speech. After intensive efforts by Keith and Cathy Hayes to teach vocal English to a young chimp, Vicki, who shared their home for several years, she could say only four words: cup, up, Mama and Papa and then only with very poor pronunciation.[4]

We now know that the block to such speech is structural. The chimpanzee's vocal apparatus is simply not capable of producing several of our vowels and his tongue is insufficiently mobile to perform many of the requirements of human speech. In consequence, studies of language capacities of chimpanzees moved to the use of visual signals, and the chimpanzee Sarah was able to communicate with humans by rearranging plastic cards of different shapes and colours into meaningful sentences. The results were far more successful than the vocal experiments and Sarah eventually learned to use a vocabulary of about one hundred and thirty terms with high accuracy.[5]

Another chimp, Lana, learned to communicate with a computer, or through her computer to humans, by signalling appropriate commands on a push-button keyboard arranged with meaningful lexograms in a simple language.[6] Signals between Lana and the computer or human experimenters were projected on to a screen so that Lana could read her instructions and also what she had herself written. Lana showed a tremendous grasp of her complex set-up, fussily correcting signals that contained grammatical errors of word order and erasing nonsense signals completely. Within a few months she had mastered seventy-five different symbols and showed a very accurate appreciation for grammar and word syntax. She could ask for food, drink, movies, windows opened, etc., correctly and could hold long discussions with her computer, or through it with her human researchers.

Even Lana's communicative abilities have been outshone by the chimps that have learned to talk by means of Ameslan, the American sign language for deaf people.[7] Experiments were started by Allen and Beatrice Gardner with a one-year-old female chimp called Washoe. Washoe quickly learned to copy the hand signs and movements involved in Ameslan, and became so proficient at this means of communication that the Gardners were often hard pressed to keep up with her flow of chatter. Within four years of first using Ameslan Washoe had acquired a working vocabulary of over one hundred and thirty signs. She showed a considerable appreciation of syntax but also great inventiveness. When confronted with novel objects for which she had no name she invented names of her own from her existing vocabulary. Thus radishes became *cry-hurt-food*, oranges *smell-fruit*, brazil nuts *rock-berries*, and melons *drink-fruit*. She also had the confidence to make her own changes in grammar. Washoe had been taught the sign *dirty* as a noun to denote faeces but one day she referred to a rhesus monkey with whom she had just had a fight as a *dirty-monkey* and soon started using *dirty* as an adjective to describe anyone with whom she was annoyed.

In 1970 Washoe moved to Oklahoma to join several other chimps being taught Ameslan by psychologist Roger Fouts.[8] Washoe continued to improve her vocabulary while helping the others to learn the sign language. Eventually Roger Fouts had four chimpanzees that could talk not only to him and other humans but held quite detailed conversations among themselves in Ameslan, discussing such important chimp topics as food and games. One chimp, Ally, even learned Ameslan signs for words of spoken English which he already understood. Thus, without being shown a spoon he learned the correct Ameslan sign for it and was able to correctly give the appropriate spoon sign when he was eventually shown the real object.

The experiments continue and the chimps reveal more and

more of their impressive faculties. One day someone will train a chimp to draw out his own lexogram symbols, one day we might even be able to ask a captive chimp about his wild childhood experiences. Already we can see how chimps discriminate and classify the world around them and we can begin to ask them questions about their own world. One of the Ameslan chimps had been trained to answer 'yes' or 'no' to the question: 'is this object the same as that one?' Thus, when shown two different pictures of cats the chimp answered 'yes', if shown a cat and a dog the answer was 'no'. When confronted with pairs of recorded sounds of wild chimps at Gombe, the chimp baffled his investigators by giving double answers 'yes yes', 'yes no', 'no no' or 'no yes'. To the chimp the question was ambiguous. He was answering two questions: 'are those two the same sort of noise?' and also the possible question: 'are those two noises made by the same chimpanzee?' Not only did the chimp classify wild chimp vocalizations in much the same way as human researchers did, but he could also distinguish between the voices of different animals although he had never met any of them!

What relevance does all this language ability have to real wild-living chimpanzees? Wild chimps do not talk, they give stereotyped gestures and vocalizations that reflect their mood and may enable other animals to better predict how the performer will behave or react in a given immediate context. Chimpanzees are unable to actually exchange useful information with one another, or are they?

American psychologist Emil Menzel conducted a series of experiments to examine chimpanzee leadership abilities.[9] Captive chimpanzees were released into a large four-acre enclosure, the layout of which was varied between experiments. A single 'leader' chimpanzee was taken round the enclosure to see the new set-up, then returned to the closed chimp house before the whole group were let out. Although different chimps varied a great deal, the whole party usually had a pretty good idea of what to expect and could set off in the right direction to investigate novel objects or find food, often running ahead of the 'leader', anticipating where he was going and trying to get there first. One smart chimp learned to 'lie' when he was 'leader', fooling the others to go off in the wrong direction before he doubled back to find the reward himself.

One of the most striking observations came when a leader chimp was shown a dead snake in the enclosure, but the snake was removed after his return to the chimp house before all the animals were let out. As soon as the door of their house was opened the chimps came out in a tight bunch. They were excited and slightly frightened, their hair was erect and they all faced towards where the snake had lain. They advanced cautiously to the very spot where the snake had been and one animal, not the chimpanzee who had seen the snake, picked up a stick and threw it down at the exact place where the snake had lain.

The 'leader' chimp may not have intentionally explained precisely what he had seen. But it was quite plain from the other chimps' behaviour that before they were released they already knew not only that there was a small frightening creature in their enclosure, but exactly where it was.

Chimps are clearly sharp to make deductions from minute differences in posture, the lie of the hair, facial gestures and bodily movements. They can deliberately control some of these movements themselves to give misleading information to their fellows and some of their behaviour looks uncomfortably like direct communication.

Higher dependence on culturally acquired technology by Early Man placed an even greater selective premium on such communicative abilities. Co-operative hunting techniques also depend on communication if the right individuals are to effectively perform their own small roles at the correct moment to achieve success. 'No, lad, you're doing it all wrong. Hold the axe in your left hand and rub it like this.' Or, 'Now you chaps, wait on top of that hill till we give the signal, then spread out and rush down the slope, driving the herd towards the stream where we will be waiting hidden with our spears.' Many researchers would argue that it was the need to give such complicated and precise instructions as these that gave the evolutionary pressure for the evolution of language. I am not so sure. Technology can be passed on perfectly efficiently with a single gesture indicating 'watch this'. I have myself, both in Africa and Asia, had occasion to order and supervise complicated constructions made by people with whom I had no shared language, yet we managed with no great difficulty. Similarly in the hunt the practices of a particular community will only employ a few variable tactics. Each member of the hunt will know from the layout which strategy to employ and will already be familiar with his own part in the action. The co-operative hunting of wolves, wild dogs, cheetahs and lions is remarkably efficient, yet these animals use a minimum of detailed communication. Certainly there are advantages for technological communication and hunting co-operation in possessing a precise verbal language, but were these advantages alone sufficient to promote the evolution of so unique a feature as human speech? I think it unlikely and indeed the bipedal hunting-gathering stage of proto-human development was achieved when our ancestors still had brains of only ape dimensions. The speedy doubling of brain size which must be associated with the development of linguistics comes much later than this. It corresponds with the period when Early Man was exterminating or out-competing the less advanced hominid species.

It became of paramount importance for us to outwit, outnumber and out-compete, both ecologically and culturally, neighbouring groups of hominids and, where feasible, to physically dominate or exterminate them. As our neighbours became more crafty and culturally more advanced, we had to make parallel advances to maintain the balance of power. The process led to a spiral of escalating intelligence and war technology which is still obvious today in the nuclear arms stockpiling and the frightening build-up of conventional armaments across all disputed borders.

It is within this context of intra-species competition and aggression that we can see the real advantages of linguistic development. It also explains the diversification and instability of all languages, for the need to give false information, confusing information or to conceal information has been as important in the evolutionary context as has been the need to convey clear and accurate details. The need to lie is one of the ape's main weapons in intra-specific competition, as my old friend the chimpanzee Figan was wont to demonstrate. Figan was one of the few select chimpanzees who could open the banana boxes at the Gombe Field station. It was unwise for him to open a box when there were socially dominant animals present for not only might Figan get attacked, he would certainly lose all his hard-earned fruits. Figan solved the problem by perfecting the acted lie. He would get up and stride purposefully from camp, so purposefully that other chimps, suspecting that he knew of a good food source, would amble after him. After a couple of hundred yards, Figan would double back alone to the deserted camp, open a banana box and enjoy his fruit in peace. The other chimps would not return; they had just been in camp and knew there was no food there, or at least only locked in boxes.

Early Man's hunting also depended on guile and false presentation: the ambush, the covered pit-fall trap and many other ruses. The savannah wildlife, that had formerly been so tame, so trusting, soon learned to recognize the upright biped as not only a carnivore but the most dangerous carnivore of all; they became wily, forcing him on to ever greater cunning. The use of intra-specific deceit in competition must have come naturally to competitive man, and speech was a superb medium in which to operate such guile. The syntactical and grammatical complexity of language is designed to conceal meaning from outsiders and restrict comprehension of the message to an exclusive few. Languages that are really designed for open communication became simplified in syntax, vocabulary and grammar, as in pidgin English and colloquial forms of Swahili and Malay. These differ greatly from the complexity of local dialects that are really forms of private code. Fierce intra-specific competition means that all men are your enemies except your offspring who carry your genes and your close kin who also share similar genetic material. Less closely-related members of the same social group or community may act in mutual co-operation against even more distantly-related individuals of other social groups and communities. Larger military alliances develop in ever wider circles. Language follows the pattern of these social relationships, so that your own use of language is only fully understood by those who know you well, your own social peers and relatives. Your use of language must be sufficiently similar to that of everyone else within your own social group to permit a certain level of communicative clarity, but with increasing social distance the information will become increasingly meaningless and eventually useless. The less your enemy knows about you, your relations, your plans and your past and present doings, the less easily he can out-compete you. Like codes that can be broken, languages must diversify and constantly change, new in-words are invented to keep the circle of in-members as exclusive as possible.

Language permits us to hide our devious underlying thoughts, we can appear friendly until we are in a strong enough position to reveal our hostility. We can make those promises we never mean to keep, pay the respects we do not feel, exaggerate or belittle our own and others' achievements, raise false hopes or allay valid suspicions. Language also enables us to play out many of our aggressive conflicts without recourse to violence, which could be dangerous among creatures equipped with such lethal weaponry as ours. Moreover at the community level it enables different individuals to put forward their points and agree upon appropriate judgements, something that was vital to the advancement of human brains over the brawn which tends to dictate the hierarchies of monkeys and apes.

Speech could have developed along either of two possible channels, the audible signal or the visual signal. For the open-country hunter the visual signal has the advantages of silence and effectiveness over long distance, but it does require complete attention between communicants. On the other hand, calls are easily understood by individuals who are concealed from one another and who at the same time are attending to other matters such as feeding, tool-making or the like. We have already seen what a great potential chimpanzees show for speech using visual signals. The vocal apparatus of chimpanzees does not have sufficient tonal variation or motor control for the production of phonetic speech, but as a result of our seed-eating adaptations the vocal apparatus of the hominids was quite different. Our rounded palate and broad, fleshy and very muscular tongue equipped man for tonal and phonetic control, and the acquisition of one of our greatest and most unique specializations, phonetic speech.

What is human speech? And how does it differ from the communications of apes? We still know very little about the way that information is programmed and arranged by the brain. Scientists are constantly altering their picture of how they think the brain works but many current researchers are of the opinion that in our brains we actually make three-dimensional, neuro-cellular maps of the spatial and temporal environment we live in, the places we know and people we have met. Thus for any concept in our mind, such as car, there is a locus in our brain, a locus that is the meeting place of all car-related ideas, such as car components, how a car works, uses of a car, appearance of cars, details of our own particular car, dangers of cars, accidents we may have had, alternative forms of transport or similar mechanical principles in other machines. In effect we make our own models of the world we live in and how it behaves; all the decisions we make are referred to our subconscious template, to check out the likelihood of various possible consequences. Psychiatrists try to probe the thought connections of our brain by asking us: 'What is the first thing to come to your mind when you hear the word dog? train? etc., etc.,' and from our answers they can get some idea of the way individuals have programmed their brainspace, whether neatly or untidily and particularly whether they reveal excessive fixations – thought connections that have become unhealthily strong. Human language is simply a code of labels overlying the basic template and if we learn more than one language we may have several layers of these labels. Sometimes we think in terms of these learnt labels but this is not essential and most of the time our thoughts are not organized in a verbal or grammatical form.

The ape's brain must also contain a basic template or working model against which to compare real life as perceived through the senses of touch, smell, taste, vision and hearing. With less brain volume available for programming and probably fewer concepts encountered for appropriate filing, the ape's mental template is probably simpler than our own but we can infer that it must be qualitatively the same. The chimpanzee also has loci for the different objects and ideas that he has experienced. When he sees a cup a chimp might not register a label such as our word cup, but he recognizes the object for what it is, how it can and cannot be used and he recalls any particular incidents he has experienced involving cups. The chimpanzee may even have a mental label for the cup concept, but unlike human language the label is not transferable and cannot be displayed for other chimpanzees to recognize, unless he has been taught by a human how to display visual thought labels. The chimpanzee shows that he has the right brain morphology for the evolution of language, though he lacks the vocal apparatus and sufficient evolutionary pressure to develop his capacity very far. His remarkable ability to communicate using sign language or to punch keyboard characters makes it dangerous to pontificate about the uniqueness of human language, but this does remain the most significant difference between human and ape culture. Men can share parallel thought pathways because they monitor each other's thoughts through spoken labels; apes in nature do not and it is this single and unique ability of man that has facilitated the incredible advances in his technology. Nevertheless it must be pointed out that man was only able to learn speech because of his ape heritage, and his language and culture are exaggerated projections of ape features rather than something uniquely human.

Writing is the logical development for a creature that communicates with audible thought labels but whose culture is based on manual artefacts. The ability to make permanent visual records has enormous advantages for the advancement of culture. Indeed it is so logical a development from speech that it was not stumbled on by accident or copied from other animals like most other human innovations, but actually planned and independently developed several times. Man needed to keep written accounts of numbers and dates, needed to make lasting testimony of his battles and hunting triumphs. He sought out ways of doing this. The Incas had no writing but recognized the need for documentation. They invented the 'quipu', a bundle of strings with knots tied in special places as lasting mnemonic records.

We can trace man's need to make visual representations of his mental concepts from the earliest cave paintings of hunting scenes and negative hand prints, through the pictorial writing and hieroglyphics of the first written languages, to the symbolic character writings and eventual phonetic alphabet writings of the cuneiform tablets, Phoenician, Arabic, Graeco-Roman and Chinese scripts. The ape's propensities for writing have never been properly investigated. We know from the abilities of the chimp Lana that apes can learn to read in symbols and arrange or type out symbols to make their own message. I don't think it would be very difficult to train a chimpanzee to actually trace the appropriate symbols and eventually to learn to draw them itself.

Writing has enabled cultural information to be spread over a wider area and longer time span than could ever have been done by word of mouth. The individual is no longer limited to deriving his cultural knowledge from his own experiences, and those of his relatives and acquaintances; he can now obtain benefit from the sum written experiences of whole cultures and many preceding generations. We have access to more knowledge in a library than we could ever learn or remember for ourselves and this has enabled our cultural and technological advance to move at such an astonishing speed.

As seed- and root-gatherers returning to regularly-used base camps or permanent homes, we were bound to discover agriculture by noticing that our favourite food plants sprang up on our doorstep wherever we spilled our food or discarded scraps. Similar primitive gardening is still seen among Australian aborigines and the Semang aborigines of Malaya, who tend rough patches of wild yams, but are still essentially hunter-gathering people. Such behaviour makes a convenient stepping-stone between the unconsicous agriculture of gorillas, who by their own activities spread and encourage their own favoured foods, and the well-planned agriculture of most modern societies.

Strong social pressures, however, have restricted agricultural advance in almost all hunter-gatherer peoples. The male-dominated hunting activity was always given higher status and importance than the female-dominated plant-gathering, even though it was the latter that provided the bulk of total diet. It was their hunting needs that forced hunter-gatherer bands to be small, and home sites to be temporary. Agriculture could only develop where people occupied sites long enough to harvest the crops they had set. Time and again the womenfolk's agricultural experiments must have been thwarted by male insistence that camp be moved again.

The most sedentary groups of hunter-gatherers were those that relied more on fishing for their protein than on hunting large game. Man always needed to live near fresh water, and the lakes, rivers and estuaries where he made his home also provided a wealth of excellent fish and shell-fish. Man's upright form and streamlined sinuous shape equipped him for wading in deep water and learning to swim. At first fishing was simply a form of hunting in water, using the same spears, bows and arrows which were used to hunt land mammals, but gradually there arose a great fishing technology with baited lines and bone hooks, nets and rattan fish traps, the use of plant drugs and poisons, stone dams to act as fish traps and development of techniques like using burning torches to attract fish at night. These waterways were not a hazardless paradise, however; large crocodiles took an annual toll of human lives. On small lakes crocodiles could be exterminated by raiding their egg caches and by clubbing sleeping animals to death at night, a feat which the late David Roberts of Lake Baringo in Kenya used to demonstrate to sceptical house-guests. Along the riverways and estuaries crocodiles were reduced in numbers but never exterminated; they remain to this day, a continual hazard and a healthy selective pressure to maintain high standards of swimming and boating.

Few tribes today still lead a harvesting way of life, relying very heavily on the abundance of one natural food and gathering it in much the same way as agricultural peoples do their cultivated crops. The Sioux Indians harvest wild rice, the aborigines of eastern Australia harvest nardoo seeds, while the Polynesians collect coconuts and Melanesians feast on wild sago. From harvesting nature's bounty it seems but a short step to developing one's own agricultural system, yet this step has been taken by only a minority of the hundreds of different local cultures of pre-agriculturalists. It is obvious why; one does not plant in one's garden plants that are available for the picking in the countryside all around. Preparing the soil, planting and caring for a crop, harvesting it and then storing the fruits of one's labours involve much more work than simply exploiting the countryside, collecting whatever foods you want from the seasonal selection. The Zairoise pygmies are surrounded by agricultural tribes, who culturally dominate them and whose languages they have learnt, but the forest pygmies have not the slightest desire to learn the labours of their neighbours' agriculture.

It used to be imagined that hunter-gatherers lived on the very edge of the subsistence level but anthropologists have been amazed at the relative affluence of these people. Even in semi-desert lands, where agricultural man could not possibly find a living, Kalahari Bushmen maintain themselves in all the style they demand on about two hours 'work' a day.[1] Hunter-gatherers generally work far shorter hours, eat better and more regularly throughout the year and show less nutritional ailments than their agricultural neighbours. Many such tribes have no word for, nor concept of, famine. Not only do agricultural peoples work harder but the fruits of their labour are not guaranteed. In some years the weather is inclement, in others the crops are diseased. The agriculturalist relies so heavily on single crops that these failures mean real hardship and shortage. The hunter-gatherer exploits many plants and animals; if one is in short supply he takes more of other foods. So why did some men change to agriculture and an uncertain food supply come about? Probably because it was forced upon them. Nature's bounty is limited. Wherever men were successful in finding food, but not numerically restrained by disease, war or the ever-present crocodiles, they would have to disperse so as not to over-exploit local resources. Often, however, they were hemmed in by geographical barriers or neighbouring human groups. Where there were no gaps to fill they were forced either to limit the numerical growth of the group to within the carrying capacity of the environment, or they could artificially increase the productivity of the environment through agricultural practice.

The change to agriculture enabled man to live at higher densities, but the cost was great: an increase in his workload and an overall drop in the standard of living. In practice the higher densities of the agricultural peoples have enabled them to numerically swamp their neighbours and then expand their domain anyway.

Under favourable conditions agriculture can produce more food than is required by the labour force alone. It is this surplus which has allowed the establishment of the professional warrior class on the one hand and urbanization, and thence industrialization, on the other. It should be pointed out, however, that there is no such thing as an urban civilization. All civilizations still depend on the agricultural plateau that supports them.

In simple agricultural communities commerce is not essential. Harvest from the community fields is shared out according to traditional rules and necessities are acquired by barter. The enormous division of labour seen in urban societies, however, is far too complex to be controlled by an effective barter system and the support of a military class demands an effective system of conscription, and/or a system of taxation. Currency was needed and developed in the form of coin, cowries, feathers or paper credit. Some quite bizarre forms of commerce still exist in remote cultural backwaters. Among the Yap islanders of Micronesia menfolk own enormous stone wheels, fei, cut from aragonite.[10] The wheels are so heavy that they are rarely moved but they are initialled by new owners after business transactions. Credit will be given against portions of these stone fei, and if more fei are required they can be cut and transported from the Palau islands several hundred miles away. The labour is so great that it acts as an efficient control on inflation. Only the really needy will undertake the journey, and there is little temptation to heap up many more fei than one is likely to require for business needs. The stone wheels are only used for important male dealings; trivial day-to-day expenses are paid by the womenfolk with their own shell currency.

There is nothing in the behaviour of wild apes that parallels barter or commerce. The practice of sharing can be recognized in mother-infant relationships of all species, and among chimpanzees who share meat after a large kill. Those chimpanzees without meat beg from those possessing it by holding out a hand or extending the lower lip to catch any discarded morsels that may fall from the feeding animal's mouth. In captivity apes readily learn to perform particular actions for food rewards, but this is probably not an example of barter but simply another demonstration of the animals' learning ability.

Captive apes do, however, have a keen sense of comparative values and I am sure that it would be quite easy to encourage the great apes to organize their life-styles on a commercial basis.

Other influences in our culture, which at first glance seem uniquely human, are magic, superstition and religion. These are common to both primitive human communities and our own technological society. How many of us do not think twice before walking under a ladder, cast a half-sceptical eye at our daily horoscope or resort to prayer when all else has failed? But are these really traits of man alone? No; again we can find the roots of such behaviour in our ape cousins.

On the whole, adult apes are suspicious, highly conservative creatures. If a particular behaviour pattern pays they stick rigidly to it and don't bother with experimentation. One day I was watching old Flo catching termites. She caught only a few termites from one particular hole and moved on to another. I went to the hole she had abandoned and by using a much longer grass stem I managed to catch quite a lot of termites, certainly more than Flo had been getting. Why was it that Flo, who must have been catching termites for over thirty years, hadn't tried a longer stem? The reason, of course, is simple. Flo did not understand what goes on in a termite mound or why it was she caught termites when she performed her fishing behaviour. She had simply learned to perform a particular cultural pattern and had learned to expect a particular reward. Her technique was probably effective most of the time, but it had become so stereotyped that she could not vary it when conditions were different.

What Flo thought as she fished for termites we cannot know, but her repeated catches must always have remained a bit of a mystery to her. In effect, her performance was very similar to some magic ritual that resulted in summoning termites from out of the bare earth, and just like magic rituals in human societies the emphasis was on the precise repetition of a long-perfected formula. To Flo there was probably no logical connection between her actions and the resulting catch, and this must be even more true of the captive chimp Lana punching commands into her computer and getting her appropriate rewards. It is precisely this lack of logical connection that characterizes the behaviour in man that we label as superstition and magic, the two foundation stones of primitive religion.

Man emerged as a highly intelligent animal and most of his daily activities were performed in a reasoned and logical manner, but all around him the world was full of unanswerable mysteries. Man hates mysteries. He has to have some answer, however unlikely, rather than face the reality of his own ignorance and inadequacy. Just as the Gombe chimpanzees were uneasy about sitting under the veranda, wondering why they were not getting wet in the rain, so man was uneasy about many of the natural phenomena he could not explain.

Unlike the chimpanzee, man could talk about his problems and discuss the mysteries. Religion provided the easy answers that made the world inhabitable, less terrifying. Volcanoes became less terrible because they were the homes of gods – gods that could be appeased by appropriate ritual. The nature of man, the reasons for life and growth, the nature of the sun, the moon, the seasons, all was explained and man's imagination was so egocentric that his religion helped him to walk tall and proud, superior to all other living creatures.

It was the evolutionary pay-off of increased self-confidence that also fostered man's indulgence in magical ceremony. If you believe that painting a picture of your intended prey on some sacred cave wall will actually help you in the forthcoming hunt, it probably will. Confidence is an important ingredient of success. I saw ample evidence of this sort of magic in Borneo where my Dusun helpers hunted the wild bearded pigs with their dogs and spears. Sometimes they would run into a bad patch, and after several luckless outings both men and dogs would get dispirited. What was needed was a little magic. New spear shafts were cut in the forest, hardened in the campfire and fitted with freshly sharpened blades. The dogs were caught and chicken feathers were burned under their noses. Next day the dogs and men would set out with renewed confidence and keenness, and almost invariably came home with meat.

We all have similar foibles, whether it is changing to a new ball when we're playing badly at golf, wearing a special tie to an interview, or carrying a good luck charm into an important examination.

Bird augury is just as important today to the Iban Dyaks of Borneo as it once was in classical Greece and Rome.[11] The Ibans regard several birds as messengers for the highest lord, Singalang Burong, who sometimes appears himself in the form of a Brahminy kite. If one of his messenger birds is encountered in inauspicious circumstances it is taken as a warning to abandon or change plan and proceed with caution. Before any important event, like preparing a new rice field or setting out on a head-hunting raid, the Ibans actively seek out the right omens and each must appear in the correct complicated sequence. It might take weeks before the Ibans are satisfied that the time is right and the gods are smiling on their project. The warring party sets off with impatience and confidence, and is usually successful, as there is no one more difficult to defeat in battle than the man who believes he is invincible. Of all the Bornean tribes the Ibans have developed the most complicated and highly ritualized lore and this has played a major role in their aggressive and bloodthirsty expansion.

Such confidence-giving rituals were not unknown among the Gombe chimpanzees. Mike discovered the 'magic' of paraffin cans; he learned how to roll them noisily ahead of him during his displays. Other chimps were intimidated by this noisy novelty, and Mike's confidence increased to the point where he not only dared to challenge the mighty Goliath for dominant position in the community but won. Mike was not a large chimp; he always looked worried and nervous and enjoyed little support from the other males. Yet, unlikely candidate that he was, he held top rank for eight years. His 'magic' talisman gave him self-confidence. Like the Iban who has performed his rituals, Mike was invincible. Once three adult males joined forces to attack Mike; they chased him and cornered him up a palm tree. This surely must be the end of his reign. Did he show a fear face and make submissive gestures of defeat as all logic demanded? No; he charged straight at his assailants who, faced with such unreasonable behaviour, scattered and fled, screaming with fear.

It was metal cans that helped Mike gain his status in the first place and he seemed to develop a cult around resonant metal objects; he couldn't pass a metal drum without beating on it. He announced his arrival in camp by slapping the corrugated iron walls of our house, and our forty-gallon rain-water butt was a special favourite. Mike would rip this drum from its moorings, beat it round camp and end his display by rolling it down the hill to the stream. Exasperated, we anchored the drum with a rock and concrete wall but it still remained Mike's totem. All his displays ended with a slapping session on 'Mike's drum'. He guarded this drum jealously and anyone who dared approach or violate the sacred totem was punished with a flying rock or a savage attack. Is such behaviour so different from the high priest who protects his gods and punishes trespassers who violate the sacred ground, or the man who imagines that he will be drained of his virility if his hair is cut?

Music is another human feature that we might imagine to be unique among the primates. But the very structure of human music is in fact clearly ape. Music is composed out of rhythmical series of discrete sound units or chords which vary in pitch, tempo, tone and volume. It is precisely such structure that characterizes all the spacing and locator calls of apes and which is quite different from the form of Old World monkey vocalizations. The territorial calls of gibbons and siamangs, the aggressive long-calls of male orang-utans, the contact maintaining pant hoots of chimpanzees and the warning hoot series of silverback male gorillas all show this form and all act as locator calls advertising the precise whereabouts of the caller. Why are all these calls and human music so rhythmical? Why do they share such similarities in form, building up with rising pitch and speeding tempo to an excited climax then dying away in a gradual tail-off? The pattern has an almost sexual form and indeed it is the rhythmical sexual grunts and cries of copulating apes that are most similar in structure to these elaborate spacing calls. The similarity may well be more than coincidental. The act of mating is the most basic of all territorial threats. The participants are unwittingly announcing that they intend to breed, to expand and to compete for food and space. We have seen in the gibbon how the first calling of a new male-female partnership stimulated neighbouring gibbons to attack the new couple. The message in that first territorial duet was exactly the same as if they had been heard copulating. It stated that a new breeding partnership was being formed.

Among orang-utans, chimpanzees and gorillas also, copulating pairs elicit great interest and apparent jealousy from nearby individuals. The mating female's offspring are the animals most seriously threatened by the potential conception and it is usually they who interfere with the copulating pair, but other animals may also intervene, for every animal's reproductive success is a real threat to all other members of the group.

I have already argued that the ancestral apes must have had a very loose social organization, much like that of modern orang-utans. Among such animals temporary breeding consortships would have constituted the greatest spatial threat, and rhythmical mating cries would have been the only evidence of this threat detectable by other individuals in the shady forest. As social complexity developed and the early apes needed to advertise their rights to breeding space more clearly, these mating cries, which already had some 'territorial' meaning, were the obvious starting point from which new social signals could have evolved. The ape 'territorial' call became an exaggeration of the original behaviour, longer, louder and more clearly formed, but still a recognizable derivative of the sexual rhythmical groans and cries.

Among the lesser apes the 'territorial' context was developed into truly exclusive spatial defence, and the rhythmical call has remained with the female accompanied by male back-up. In the great apes, however, 'territorial' roles became features of the male sex and it is they who produce the rhythmical spacing and locator calls. Each species has modified its common behavioural heritage in rather different ways to suit its own divergent social needs, and in man the rhythmical beat has become readapted in several directions at the same time.

Human music has been retained for its conservative territorial function. The military march and the tribal war dance are classical territorial threats – audible and intimidating to the opposition but binding and encouraging to the performers. The addition of percussion instruments is no accidental innovation. During his long-call the male orang-utan sways vegetation as he calls and for extra effect breaks off dead branches which crash noisily to the ground. Gorillas reinforce their hoot series with rhythmical chest-beating and vegetation-throwing. Chimps occasionally beat their chests in display, but more often drum on resonant tree buttresses and throw rocks and vegetation. The human warrior dancing to the pounding beat shakes himself and his weapons in a menacing display. Just like the ape calls, the human war dance works up to a pseudo-orgasmic climax with the performers reaching a frenzied peak of excitement and aggressive arousal.

Music has also had a role to play in peaceful community-binding ceremonies and feasts, religious rites and pure aesthetic performances, but perhaps the most amusing sideline of all is the development of the pop culture. Pop music is an essential part of our pattern of courtship; it is designed to promote the partner-discovering dance behaviour of young unmarried adults. Pop music is nothing new. Amazingly similar music, the same driving sexual beat, is used to accompany the oldest African tribal dances, themselves a part of ancient courtship patterns. The evolutionary wheel has turned full circle. The rhythmical call that arose from sexual emotions in the ancestral ape has become modified and redirected. It has served as a territorial warning, as a community-binding chorus, and now, in modern society, it serves to re-arouse those sexual emotions from which it first originated.

We have already seen the buds of aesthetic taste in our cousin the chimpanzee. Chimps dress up and enjoy self-decoration; they have a natural sense of rhythm and love to beat on drums; they make wind-borne patterns with dust, scratch in the mud, and execute splendidly colourful paintings.

I have argued that these are the inevitable by-products of an animal which requires great acuity and appreciation for visual and vocal pattern and which is also a playful, inventive, inquisitive, cultural creature. It is young chimps who are playful and inquisitive, it is young chimps who show the buds of aestheticism. In the wild such behaviour, like Fifi playing with dust clouds and Goblin scratching in mud, take up very little time or effort. Under captive conditions, however, these creative faculties can be developed to a higher level. Zoo chimpanzees love painting for no other reward than the pleasure of making colourful patterns. Painting in chimpanzees develops along similar lines as seen in human children, by means of experimental manipulation. Different types of scribbling develop into advanced and well-controlled pattern arrangements. Indeed, as Desmond Morris noticed in his detailed studies of ape art,[12] 'where compositional factors are concerned this control is more active and better organized in the apes than it is in the young human.' Morris also describes how the art of one chimpanzee, who was offered food rewards for his work, deteriorated steadily in design, balance and composition. The artist was degraded into a commercial craftsman, a pattern not unknown in our own creative species.

Man has taken these aesthetics much further than any ape and devotes a great deal of his energies to artistic expression, not just in his youth but throughout his adulthood as well. Why do we expend so much energy and money on building splendid structures, carving beautiful and evocative shapes, painting, singing, acting and writing in prose and poem? Are these not diversions from nature's goals of achieving reproduction? Are these not wasteful activities that should have been selected against by both natural and artificial selection? Since they are so universal among our species, the answer is obviously no. Art must be adaptive, must in some subtle way help the individual to leave more offspring, or offspring with better chances of reproducing in their turn. Why then are we so artistic? It is not sufficient to say that the craftsman gains pleasure from his work or that the less-talented benefits second-hand by viewing or owning the craftsman's art. We must not fall into the trap of mixing motivation with evolutionary adaptiveness. If a behaviour is adaptive, i.e., helps the individual to achieve evolutionary success, then selection will find an appropriate motivational pathway to ensure that the behaviour occurs. Of what evolutionary benefit can artistic expression be to us?

Social pressures have certainly helped to foster some forms of art to glorify individuals, nations, capitals, religions, etc. Those in power breathe like everyone else, they eat, they sleep, play and breed like the rest of us. To really demonstrate their greatness they have needed to show their wealth, or spend it on those treasures which are always appreciated in cultures with an oligarchical power structure. Social pressures, however, have merely exploited something already deep-rooted in our species.

As a species we spend a great deal of time in simply learning or acquiring the culture that has already been established by our forebears. The chance of original discovery is slight at the juvenile stage that is so inventive in chimpanzees. In man there has been good reason to prolong childish inquisitiveness into adulthood. Without this extension of the inquisitive period we might have stagnated in cultural conservatism.

Secondly, there has been a need to keep our mind as well exercised as our body. It is not often that we are called upon to make lightning decisions, but if the occasions arise we must be capable of acting fast and clearly. The brain must be able to swing instantly into fourth gear. It must be kept well tuned for such important moments, and art and intellectual stimulation are ways by which we keep it tuned.

Art also serves a recreational function. It is a pleasurable activity. To remain healthy the body and mind must be fulfilled and satisfied. A happy body is a healthy body. Life must be fun, worth living. Collecting seeds and tubers is not a very fulfilling or exacting task for an active animal like a man and it has been advantageous, therefore, to make our leisure periods as interesting as possible. We have developed an enormous repertoire of recreational pastimes to satisfy this need, with the arts playing an important role.

Art also serves to boost our ego as a species. It is part of man's domination of nature, it enables us to walk tall and confident through a world that would otherwise appear hostile and overpowering. Art is man's proof of his own divinity (or divine ancestry), the proof that he is not as other beasts.

Perhaps most important of all, however, is man's desperate need for self-expression. Man is an ape, a haughty individualistic creature, not the product of some long lineage of zombified social animals. Man's own potential danger to himself has forced him into a highly social way of life. But in becoming social each individual finds himself in danger of being swamped as a mere insignificant unit of a larger social community. On a smaller scale the chimpanzee faces the same problem, and reacts in a similar fashion. Every time a male chimp joins up with a party of his fellows, he performs his own dramatic arrival display. He cannot afford to be taken for granted; he must enforce his own personality at every social opportunity.

Man, too, must make his individual mark, but he is often encountering strangers and cannot afford the latent threat of an arrival display; instead he wears his individual display on his person. He expresses his own personality in his bodily adornments. Self-decoration is as old as man himself. In every collection of human Stone Age artefacts is found some evidence of beads or bodypaint-holders. There is no culture today that is content with nature's own shaping of the self. We must dress up, decorate ourselves, mutilate ourselves in a desperate cry to be noticed as individuals. We accord to the fashions of our race, our village, our historical period, our age group, our beliefs, for these too serve important social functions; but we never exactly fulfil those fashions and in the tiny ways we depart from them we display our own personality.

When I make a stone axe, I make it as functional as I can, as light and graceful as its use allows, but I also make it my own axe, I make it personal to myself. Either by its shape or a special mark, I must somehow distinguish it from the axes of all my peers. The craftsman too leaves his mark, although it is for someone else he makes the goods. Each piece of work carries a little of the soul of its maker and therein lies the real origin of human art. The artist's monogram is the ancestral artform and it is precisely for that reason that the fake or the imitation has always been artistically valueless. It may be as beautiful as its original model and just as well executed but it is not art because it is not the expression of the maker's personality.

The level of man's technological, linguistic and artistic achievements place him culturally far above the apes and other animals, but there have been few great discoveries or major leaps in his cultural advance. It has been a gradual process of modification and improvement, adapting old methods to new problems, recognizing the significance of accidental discoveries and making use of newly-understood properties. Ideas such as building houses or making pots may be copied from nature. African mud huts bear a remarkable resemblance to the nest cells of mud-dauber wasps, and the classical African clay water pot (with the narrow-tipped aperture) is an exact replica of the nests of the potter wasps. Threads and nets are used by spiders; tailor birds sew their nests together. Certain African caterpillars build spiny barricades around themselves before they pupate, just like the thornbush fences that surround African compounds. Ants use leaf and twig rafts to cross pools and streams; snakes, wasps and spiders employ poisons to enhance their effectiveness at hunting. Ant lions dig pitfall traps to catch their prey, dung beetles exploit the principle of the ball or roller which preceded the invention of the wheel. Hornbills construct mud homes with windows; the list is endless. A few of our inventions may be attributed to flashes of original genius but most probably resulted from the old ape behaviour of watching, understanding and doing likewise. In our modern world submarines and aeroplanes are merely the latest examples of ideas derived from nature and modified to our own requirements. The very basis of scientific investigation is the minute examination of the chemistry, physical and biological properties of our surroundings so that we can manipulate our habitat to our best advantage.

Man's cultural advances have been shaped by his need to survive in harsh and varied habitats, shaped by his desire for greater creature comforts and shaped by his need to live at higher densities. Most importantly, however, man's cultural advances have been shaped by his need for military security.

It was military need to find superior weapons that drove men to such perfection in stonework, ironwork and bronzework. It was the military development of fortification engineering that led to progress in urban architecture. It was the utilization of chemical explosives and incendiaries that pushed forward our understanding of chemistry and physics. Military needs for fast and efficient communication have played a vital part in the development of writing, signalling, telegraphy, radio and radar. It was military drive that led to the use of horses, chariots, tanks, aeroplanes and shipping. Military support has speeded our advances into electronics, nuclear power, computerization, space travel; even biology and psychology have been helped with defence funds because of their military usefulness.

In times of economic stability and relative peace these technological advances have benefited us all, increasing our everyday efficiency and comfort, but the driving force behind our inventiveness has so often been military necessity. Even today the best way that a scientist can guarantee continued government funding for his research is by pointing out the military significance of his work.

The need to defend ourselves from, and if necessary slaughter, our fellow man has obviously been of enormous importance to our species over the past few thousand years, of sufficient importance to shape the development of our entire socio-economic organization and cultural advancement. Yet paradoxically it has been this very same need that has driven us to greater degrees of mutual co-operation and social tolerance than we find in any other ape or even monkey society.

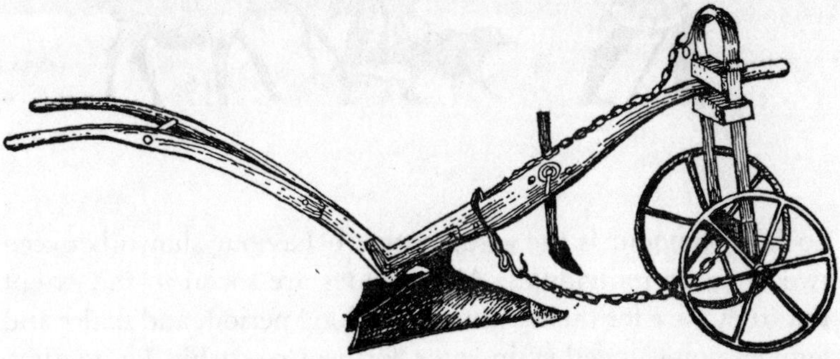

CHAPTER 9

Man and the Family

Social behaviour is the co-operative behaviour shown between two or more individuals. All primates are social to the extent that they care for their young over a long period, and males and females come together in some form of courtship for mating. Many primates go much further than this and live in social groups where they co-operate not only in breeding and rearing young but in finding food and defending the group against predators. Nowhere is such co-operative behaviour seen better than in monkey groups. The composition of these is fairly stable and the unity of each group is strengthened by cohesive social bonds between mothers and their young, between adult females and the dominant males, between different adult females, between siblings and playmates, between the resident adult males in the troop. These bonds are reinforced by shared pleasurable activities such as play, grooming, sex and feeding. Rigid social hierarchies are maintained throughout the group and determine who can do what, so that we can ascribe individual roles to the different troop members. Monkey troops may vary greatly in size and composition both within and between species, they exhibit different levels of social tolerance between the various troop members in different social contexts, but overall the structure of monkey societies is remarkably uniform.

The apes show no such consistent pattern. Ape societies are not only different from those of monkeys, they are different from one another. The lesser apes live in monogamous territorial family units. The gorillas range in larger, fairly stable foraging parties, held together by the relationship of each adult female to the silverback male. Chimpanzees live in open communities where males show unusual tolerance to one another and co-operate to preserve the security of the community range, but females lead rather individualistic and solitary lives. Orang-utans show even greater individuality and are as solitary as any mammal can be. Our own species regards itself as highly social, but again social groups are not clearly defined. Typically we recognize family units within larger and often very complex communities. As individuals we co-operate much of the time with other members of our community but our behaviour is, nevertheless, extremely individualistic.

Whilst it is impossible to characterize simply the form of ape societies, we can at least say that apes typically do not live in stable, rather closed social groups like those of monkeys.

The basic social unit common to all ape societies is the maternal family unit. Young apes enjoy a prolonged period of dependence on their mothers, and even when they are sufficiently self-sufficient for her to have another infant the mother-offspring ties are maintained. Consequently most young apes are able to establish social relationships with their siblings even though there may be several years between them. When they reach adolesence or sub-adulthood offspring break away from their maternal family units to range independently, but social relationships are usually still maintained by grooming whenever the members of the family meet up again.

All apes, and humans, typically bear one young at a time. Birth is usually at night. Apes give birth unaided and usually lick their babies clean; they eat parts of the afterbirth and chew through the umbilicus after it has begun to dry. Human mothers are usually assisted in childbirth and babies cleaned in accordance with cultural toiletry customs. The large head size of our brainy offspring has made childbirth a more difficult and painful affair than it seems to be for the other apes but, nevertheless, behaviourally we have departed very little from the ape norm. Women can, and do if need be, give birth unaided without cultural aids or toilet facilities.

Neither human babies nor ape infants are born with much limb strength or control, but young apes can grip powerfully with their hands and feet and within a few hours of birth can suspend their full weight from the mother's body hair. Human babies also are born with this powerful grip, but they have no need of this strength and lose the reflex some weeks after birth. Human mothers may have long tresses of hair, soft graspable breasts or clothing from which babies could hang but in practice these do not compensate for the loss of ape body hair. Moreover human babies are about twice the birth weight of their ape counterparts and the strain of suspension would be too great. Somewhere along our evolutionary path it became expedient for us to carry our babies rather than expect them to hang on by themselves. The gripping ability of newborn human babies is merely a vestigial reminder that our ancestors did indeed hang from their mothers' fur. Human babies are now supported by their mothers' arms or carried in some form of culturally-designed clothing, net, cloth sling or elaborate harness. We can even trace this behaviour back to our ape ancestors, for ape mothers will carry their babies if these are too weak, ill or injured to cling on for themselves.

There is a great deal of cultural variation in the way that human babies are carried, how long they are carried for and who carries them. Western industrial countries represent the extreme of limited attention, with babies being deposited for long periods in cots, prams, cradles and other warm springy containers. In other cultures babies are rarely out of physical contact, being carried either by the mother, a willing grannie or elder sibling. From earliest times our species has occupied permanent or semi-permanent campsites where babies could be safely left during restful periods in the day and throughout the night. Such behaviour is unique among the apes and rare among primates, but several nest-building lemurs and also lorises regularly park their babies whilst they search for food.

Our babies have never quite learnt to accept this neglect. They are still comforted in the same way as young apes, by the warmth of the mother's body, by her regular heartbeat and the rhythmical movements of her heaving chest. The most successful baby receptacles imitate these stimuli to lull the babies into a greater sense of security. Cradles rock rhythmically, and in many cultures babies are suspended in rocking slings or hammocks to sleep. In South-East Asia baby sarongs are suspended from strong metal springs so that as the baby stirs he rocks his cocoon and the action soothes him back to sleep. We wrap our babies in warm clothing until they can hardly move and sing them lullabies, but they are still not compensated for the lack of physical contact and they are great criers. Ape babies will cry out if they are in pain or afraid or have been left behind by their parents, but these are rare occasions. Ape babies are never far from their mother's nipple. When they are hungry they simply clamber round and suckle. Human babies cannot do this, so they have had to evolve a signal to call their mothers back. Crying has developed into a whole new concept, discordant bawling so loud and unbearable that even the most callous of mothers is soon compelled to drop whatever she is doing and attend to her neglected offspring.

Human infants cry when they are hungry, uncomfortable, in pain, sick, frightened, lonely or simply feel unloved. Our children cry much more than ape babies, but even within our own species we can recognize differences in crying levels. I have spent much time in Asian kampongs and African shambas and villages where children have no toys, no fancy childcare paraphernalia but are pampered, coddled and loved and only cry in real distress situations. In Borneo, for instance, crying is taken as a sure sign of intestinal worms. It is in Europe, where babies are materially spoilt but comparatively deprived of physical affection, that baby-crying is accepted as the norm. We are afraid of spoiling and pampering our children although we really do not know the sociological consequences of such behaviour. There is no acceptable evidence that too much affection in childhood ever hurt anyone or encouraged undesirable traits in the adult character; whilst there may well be dangers that material spoiling can lead to selfish materialism.

Human infants develop rather slower than their ape counterparts and are dependent on their mothers for a much longer time. They are slower to feed themselves, to crawl and walk, but their pattern of developing self-sufficiency is typically ape-like. Because of this long maternal dependence it is important that human young are not jeopardized by the early birth of a second infant. The mother's fecundity is reduced in two ways. Hormonal feedback from continued suckling reduces the incidence of conception, especially among pre-industrial societies where the natural birth interval is close to the three-year gap typical of wild apes. Only if the infant dies or is weaned will the mother conceive again. Infant and juvenile interference with attempted copulations also helps to maintain wide spacing of births. Human youngsters, like their chimp and orang-utan cousins, become distressed and jealous if they see their mother being mated and try to disrupt the behaviour. Where the whole family shares a single living and sleeping space, with little or no privacy, this interference by children must effectively reduce the incidence of copulation. In many primitive human societies there are additional cultural taboos against mothers copulating within a certain period of time after giving birth.

In their growing social awareness and the development of new relationships with siblings, other infants or play-peers and with other adults, whether relatives or mere neighbours, human youngsters follow traditional ape patterns. None of the apes has such a complex social system as is found in some human societies but there is nothing unique about the way our children develop socially. In fact their relationships are very similar to those seen in chimpanzee communities.

In the development of human speech and in our level of cultural and technological instruction we do show elements of social behaviour not seen in the other apes but, as we have already seen, these uniquely human qualities are entirely dependent upon, and are the logical projection of, our ape heritage.

As the young human grows up, it spends more time with its age-peers in play or instruction and less time with its mother, who may well have another infant by this time. The juvenile human, however, continues to reside with its mother's family and normally eats with the family unit. This link between the young and their mother is the deepest and most prolonged social bond of the early years and, in many individuals, of their whole life. Even when the need for independence forces the adolescent or sub-adult to leave the family dwelling place, the bonds are strong and remain so until death.

Human social interactions are much richer than those enjoyed by apes. Common language and cultural aids permit us to indulge in a far wider range of pleasurable shared activities. Social gossip does little more than replace physical grooming as the commonest form of social interaction, but we can stimulate one another's intellects in deeper discussion. Shared viewpoints, shared news items, stimulating games, whether physical, intellectual or of chance, shared fun in eating, drinking, dancing and flirtation, all give us numerous opportunities for forging strong social bonds; and perhaps strongest of all is the shared comradeship in arms against a common enemy.

One nearly universal element in human society is the father figure. Paternal behaviour is a rare feature among apes. The silverback gorilla shows concern and great tolerance for the infants in his group; he provides much of the security and leadership we associate with the paternal role but he remains aside from the maternal family unit. Male chimps, too, will protect and tolerate the group's young. Once when Melissa left without her infant Goblin, it was the dominant male Mike who picked up the baby and carried it up the hill to Melissa. On the whole, though, the adult males again remain apart from family units, and in the chimp multi-male system knowledge of fatherhood is even more tentative than in gorillas. Among bonobos and Sumatran orang-utans there appears to be more permanence to consortships and the same male may range with a female and her young for weeks or even months at a time. Even if he is not always the real father of the female's young, undoubtedly the consorting male develops a paternal-like relationship with them, but we still know little about the strength or permanence of such relationships. It is only among family-living gibbons and siamangs that the males perform a clear paternal role. In the case of the siamang the male even takes over the responsibility for carrying the infant in its second year. Thus, among all the apes, the adult male only constitutes an important and integral part of the family unit in the lesser apes and man himself. Yet the gibbons are our most distant relatives among the apes. It seems, then, that monogamous family living is not something we have inherited from a common ape ancestor; instead it has evolved separately in these two lines and for quite different reasons. In evolutionary terms both gibbon and human monogamy probably arose from the same sort of prolonged consortships that we find in the rainforest great apes, orang-utans and bonobos, but here the similarity ends. Gibbon families are self-sufficient territorial breeding units; human families are not usually self-sufficient, they are sub-units within larger social communities. Males and females can be classed in the same family unit because they sleep together, may feed together and may share the responsibilities of providing for, and instructing, their offspring, but the family units are not spatially cohesive. The males traditionally leave the family unit, perhaps for several days, to hunt or work, activities in which they participate with other males from the same social community. The womenfolk lead a more base-orientated existence, working or gathering vegetable foods with other females. In fact the organization of communities shows much greater similarity to the chimpanzee system than to lesser ape society. Moreover the nature of human courtship and the types of social bonds that hold human families together are quite different from those seen in the lesser apes.

Let us consider the human male's responsibility in providing for his offspring. He provides both in the immediate sense of his contribution towards the family cost of living, either financially or in kind, and also in building up wealth, status, land, or a commercial business which will be taken over by his offspring in their turn. From the days when man first became a hunter-gatherer, and perhaps even earlier, there has been a major division of labour between males and females in our species. Females undoubtedly make the major contribution in the direct effort of bearing, rearing, feeding and looking after the offspring, but the males make the greatest economic contribution to the family budget. In commercialized societies the father's wage packet is usually larger than the mother's. In agricultural societies the males perform the hardest physical labours, while the females are given the lighter, though often more prolonged and tedious tasks. In hunter-gatherer societies the men hunt for the more highly-prized protein and their contribution is valued higher than that of the plant-gathering women, even though the latter may contribute more to the actual energy budget. In all societies it is the males who provide physical security and in many societies this is regarded as the ultimate contribution, excluding the male from almost all other labour.

While in Zaire I was surprised by the attitude of two porters I had hired. After complaining about the weight of my luggage, lifting it up, groaning in pain and haggling about the price of their help, they finally loaded it on to their wives, who were already bent double with their own luggage and crying babies. This seemed to me to be a prime example of male chauvinism (and sexism) until I realized that the system could only have survived because it was adaptive. Those tribes who left their menfolk free for fast action in defence of the travelling party or to hunt game along the route would have a better chance of survival than those where the men shared the workload equally with the womenfolk, but were hampered by their loads if they fell into an ambush. True to tradition, my two guides carried only their bows and arrows and took a keen interest in the monkeys and other forest life around us.

Among the apes adult males contribute very little to the upbringing and welfare of the young but they do all fulfil important roles in maintaining the security of their social systems and do, by their behaviour, benefit their offspring.

In monkey societies young males move between groups, so group succession follows the female line. In the apes, however, it is the females who move into established male ranges. Succession of range tenure follows patrilinear patterns, so that males can contribute greatly to their sons' chances of social success.

Male gibbons are most active in the defence of their territories. As their families grow they require larger territories and they enlarge their range by exploiting any weaknesses shown by their neighbours. The more successful the gibbons are in their territorial efforts, the larger is their range, so that the area they occupy may, in fact, support a density of gibbons which is below the carrying capacity of the forest. When the male gibbon dies, it is his own peripheralized sons who are most likely to take over his territory and, if it is large enough, two or more sons may be able to set up breeding partnerships in the area previously occupied by their parents alone.

Gorillas do not show such territorial exclusiveness; their ranges are too vast to defend adequately and in any case range overlap permits important exchange of females, a facility which is mutually desirable to all concerned. The core areas of gorilla ranges are fairly exclusive, however, and are defended by the maze of fresh gorilla trails which discourage intrusion by another group. By maintaining as large a range as possible and by attracting as many females as he can into his group, the silverback is giving his male offspring the best possible chance to breed after him. As sub-adults these males may live on the outskirts of the group range but they return to take over the group when the silverback relinquishes his leadership because of age, illness or death. The fact that a young male has been absent from the group increases his chance of taking over the adult females since they are not used to seeing him treated as a subordinate.

The old male orang-utan continues to defend the area in which he has been sexually active long after his reproductive prime. Like the gibbon, he effectively maintains a population vacuum to prevent anyone else breeding and guarantees that his sons are not out-competed or swamped by immigrants before they are themselves sufficiently mature to call and give their own aggressive displays.

In chimpanzees the situation is somewhat different. The adult males co-operate to guarantee the security of the community range but again the succession is male to male. It is the young males who remain in the communities of their birth whilst females frequently transfer to a neighbouring community on reaching sexual maturity. Males can be seen to be protecting space for the benefit of their sons.

Why should it be the sons who benefit most from the male ape's territorial efforts? The answer is simple. Female apes are readily accepted by strangers; males are not, they are threatened. Almost all adult females in a population will breed and the difference between the most successful mothers and the least successful mothers is small. There is little a dominant male can do to improve the reproductive chances of his daughters but his sons are another matter. A really successful male could father dozens of young, while many males may leave no young at all. Anything an adult male can do to help guarantee that his sons will breed will enormously increase his own evolutionary effectiveness. It does not matter whether the male knows who his sons are or not, it does not matter whether the male helps his own offspring or whether, as in the case of the chimps, the male merely assists the community to help all the community's sons; the evolutionary function of all these status-territorial-related behaviours is the same. Every breeding male does his utmost to give his own sons the chance to reproduce. The male apes are fulfilling precisely the same function as the human father who leaves his wealth to his sons.

In chimpanzees the differentiation of roles according to sex is also marked and extends even into food collection. It is the females who are the busy insect hunters, probing stems into termite mounds or making safari ant lollipops. The bigger adult males are the great meat hunters, co-operating with one another to catch baboons, bushbuck, monkeys and young pigs. Such sexual specializations, however, could only lead to co-operative division of labour if there is some element of food sharing. Food sharing is well established, and chimpanzees have several ritualized forms of begging for food. Females sometimes hand food to their young or other nearby animals. Animals do not usually proffer it to dominant males but the latter regularly steal food from their subordinates. Males, too, show food-sharing behaviour. After a successful hunt the other males, as well as females and young, queue up to get portions of meat from the male who made the kill.

Human sexual determinism could well have its roots in a community system very like that shown by the Gombe chimpanzees, but whatever its origins the co-operative division of labour according to sex is deep-rooted in our evolution, and can still be recognized in all surviving cultures. Not only has there been biological selection for girls and boys to fulfil differing roles and therefore to possess differing adaptations, but these differences are reinforced by strong cultural and environmental pressures from the earliest stages of instruction. Little girls are brought up to be mothers. They spend more time with their mothers, they are encouraged to participate in domestic chores, to help with looking after younger babies. They are discouraged from joining the boys in the more adventurous types of play. They are continually exposed to role-enforcing images of what is right for little girls and what is right for little boys, what women should do, feel like and enjoy, and they are taught to regard the heroic father figure with due awe and respect. Little boys are trained to be the hunters, the warriors, the great providers, the more active and dominant sex. It is during our childhood pastimes that we learn the roles we are expected to assume in adulthood. In western societies where women are expected to be homely creatures girls are discouraged from fighting and rough-and-tumble games. Among Kalahari Bushmen, however, the adult female role is far more active, for the women not only collect the bulk of the group's food but travel considerable distances when the nomadic bands change campsites. Bushmen girls show more energetic play patterns than their western counterparts and indulge in almost as much fighting and rough-and-tumble games as do Bushmen boys.[1]

Juvenile play is also important in developing sociable tendencies appropriate to the society in which you grow up. This is certainly true both in apes and in man. Research on rhesus monkeys has shown the importance in early life of forming play-peer relationships. Young monkeys that successfully form relationships with their playmates develop co-operative social relationships when they become adults. Young monkeys that are naturally or artificially deprived of the opportunities to form these early bonds find it very difficult as adults to integrate into their social groups. They tend to be isolationist and over-aggressive and quickly become labelled as social outcasts by the other monkeys.

The same principle applies to apes. Because of their solitary way of life young orang-utans are naturally deprived of play-peers and become anti-social adults. However, if orang-utans are able to socialize when young, as they are in many captive collections, then they become more sociable as adults. Such social flexibility is important in the wild. For instance in areas where the forest offers particularly rich feeding, orang-utans are only able to fully capitalize if they can increase their social tolerance and live at higher densities.[2] At higher densities developing youngsters enjoy more opportunities for social play, and thus acquire appropriately higher social tolerance when they mature in their turn.

The same is probably true in human groups. The sort of social tolerance and co-operation that is required of an adult living in a densely populated, fertile, rice-growing area must be very different from that required by forest hunting peoples who co-operate in small parties, or by temperate zone farmers where individual families occupy their own smallholdings, sometimes far from their nearest neighbours. The types of childhood experience that are likely to result from these very different ways of life will all tend to reproduce similar levels of social interaction from one generation to the next.

Eventually, having attained sexual maturity and with education complete, the young men and women leave the parent family unit. The pressures for them to do so are many and various – economic, courtship and marriage, or merely the need to assert their own independence.

Courtship is an area of human behaviour about which there is much confusion and this has led to a great deal of individual unhappiness. Courtship is simply the behaviours involved in selecting, attracting, and copulating with a mating partner and in species like ours, where breeding partnerships co-operate in rearing their young, it also includes the formation and maintenance of a pair-bond. It is important to see human courtship in its proper perspective as an ape behaviour, but patterns of ape courtship are themselves obscured by superficial diversity. At first glance there seems to be no common pattern to the pair formation and sexual behaviour of the monogamous territorial gibbons, the promiscuous social matings in chimpanzee groups, the occasional pair consortships of chimpanzees away from their groups, the rather casual copulations of gorillas, the temporary consortships of sub-adult orang-utans, the calling up of females by adult male orangs, orang-utan 'rape', and the bewildering variety of courtship patterns seen in our own species, variety that forms the basis of half our written literature, many of our jokes and much of our social gossip.

We have already argued that our ancestral apes had a rather primitive orang-utan-like society where adult males and females formed temporary consortships for mating only, and where the females were left to rear their infants alone. Such a pattern is very different from that found within the regimented group structures of Old World monkeys. In monkey troops it is usual for each breeding male to establish mating rights to a harem of females with whom he ranges. The breeding male will only tolerate younger subordinate males copulating with his females when these are out of oestrus, at such times as there is no threat to himself, and the experience may be important in the development of the junior males. But when a female reaches the peak of oestrus her condition is easily recognized from unmistakable olfactory and often visual cues, such as swollen red posteriors. Then the breeding male monkey maintains exclusive mating rights, by aggression if necessary.

The traditional ape pattern of courtship has been amended in the modern apes to suit their own particular life-styles, but all are clearly variants of the ape rather than the monkey theme. In gibbons and siamangs courtship has become almost synonymous with the formation and maintenance of a territory. The initial pair formation is territorial rather than reproductive and the pair becomes bonded more tightly by each partner's need to regularly perform the territorial duet with a calling mate. Actual sexual behaviour is infrequent. During its entire life a gibbon may copulate only a few dozen times.

In the gorilla there has been selection for extending courtship into a lasting pair-bond but gorilla society is polygamous, with a large high-status male attracting a succession of different females to join his group. Like the gibbons, gorillas indulge in sex as infrequently as the slow reproductive needs of this species require. Group cohesion is maintained not by social sex but by each female's need for the security that group life offers.

Chimpanzees have developed two main sexual strategies. At the centre of community activity females can enjoy regular sexual attention from all the males; alternatively one female may pair up with a single male and go off on a temporary monogamous 'honeymoon'. The incidence of such consortships is low but a female is much more likely to become pregnant during such a honeymoon than during the mainly social sex activity of everyday community activity. It is interesting that chimpanzee and bonobo females are unique among the apes in having large pink bottoms to advertise their oestral condition, though this is common among terrestrial open-country monkeys.

The orang-utan exhibits several different reproductive strategies according to the age, status and ranging behaviour of the animals concerned. At one extreme females may be raped by strange males they encounter, at the other extreme adult females may initiate sexual behaviour with adult males to whose calls they have been attracted.

It is not easy to generalize about ape courtship, but we can say that in all species males attract females into their own range and courtship occurs during a temporary consortship which may or may not be extended into a more lasting pair-bond.

Courtship behaviour in man is typical of this ape pattern although, like the orang-utan and chimpanzee, we have several different reproductive strategies which may be adopted according to particular social conditions, in addition to several types of sexual behaviour which serve social rather than reproductive functions.

Sexual behaviour in our own species, as in other apes, begins its development early in the life of the individual. During infancy we discover our genitalia and experience our first erotic sensations. As juveniles we try to understand the sexual behaviour we see among adults in terms of our own limited knowledge of our bodies. Boys and girls indulge in 'dirty' games of physical exploration and experimentation, as do young chimps and orangs. With adolescence and the development of sexual maturity these immature games become better directed into adolescent crushes, fantasies, masturbation and increasingly daring experimentation with kissing, cuddling, fondling and ultimately our first sexual copulation. The developmental patterns are very much the same from race to race and culture to culture, though social attitudes to developing sexual experience may vary considerably.

In our own western culture brides were until recently expected to be sexually naïve but today these girls are the exception rather than the rule. In twenty years time it may well be unusual to legally marry at all. In parts of Central Europe and among Highland folk in Scotland it was common practice for a man to propose marriage only after the girl had demonstrated her fertility by becoming pregnant. Many African tribes regard adolescent sexual experiences as acceptable but the male must not introduce sperm inside the girl. In some tribes girls wear leather aprons and anything is permitted up to penetration of the apron. In other tribes the men are expected to have sufficient control to withdraw before ejaculation. Possibly such reproductively futile behaviour as anal and oral intercourse originate from such taboos although both are also found among captive orang-utans. Western society's increasingly liberal attitude towards adolescent sex is largely due to the availability of free contraception. It is the unacceptability of the heavy responsibility thrown on to unmarried mothers that has led society to make strict rules restricting early sex.

Human cultural practices of birth control are only recent improvements of natural physiological methods of controlling births. Like female chimpanzees, girls who use no contraception are quite likely to become pregnant during their very first sexual experiences or their first sexual intercourse after a long break. But, as in female chimpanzees, there seems to be some physiological mechanism by which women who are having intercourse with a succession of different males rarely become pregnant. The prostitute phenomenon has long intrigued students of human contraception. Collecting reliable data about such matters is extremely difficult, because most women are understandably reluctant to discuss their sexual habits, and those that are willing to do so may not be a fair sample. Nevertheless there is considerable anecdotal evidence to support the long-held view that prostitutes and other 'sexually loose' women become pregnant far less often than one would expect. This situation seems remarkably like that of the regularly-cycling female chimps who can have extensive promiscuous sex with large numbers of males but rarely conceive. Why this should be so we do not know – it may be due to hormonal changes or an immunological response to semen exposure. The fact remains that a female chimp's chances of becoming pregnant are enormously enhanced when sex is limited to a single partner during a consortship and presumably the same is true in our own species. Indeed, all ape consortships and pair-bonds can be interpreted as attempts by individual males to improve their reproductive chances with a given female by restricting her sexual availability to other males. Such a system has obvious advantages in that both sexes can enjoy wide sexual experience with a low risk of pregnancy, so that they have the widest choice from which to select when and with whom they actually breed. That such sexual playing of the field would be adaptive is confirmed to some extent by the fact that now that women do have almost total emancipation from unwanted pregnancies, social intolerance of pre-marital sex has become greatly relaxed and girls have been only too ready to enjoy such new-found sexual freedom.

It is from these early sexual experiences and encounters that deeper, more serious and lasting relationships develop. The classical human romance of falling 'in love' is not only the most important courtship pattern in our species but it also corresponds most closely to the traditional ape consortship. Typically the lovers are two sexually mature adults who are free from other strong social commitments and who form a strong mutual bond of physical attraction. The consortship is very similar to that seen in apes – the pair travel around together and spend every possible moment together (both day and night) that social, economic or cultural pressures will allow. Like consorting apes the courting couple seek out quiet localities away from the centre of community activity.

As in orang-utan consortships the human pair indulge in bond-strengthening activities derived from juvenile play and social grooming: hand-holding, gazing into each other's eyes, cuddling and kissing all help to strengthen the bond and make it more clearly recognizable to outsiders. The partners become mentally anaesthetized by their condition, hormone levels go wild, they lose their appetites and become inefficient at work. In fact the individuals become so absorbed in the love affair that they are oblivious to much of the world around them and blind to the faults and unsuitabilities of their partners. It is this irrational element of human love which makes it such a mysterious and difficult condition to understand, almost pathological in its effects. Although the sex drive is at a peak during such courtship, cultural pressures frequently encourage the partners to hold back these urges with the promise that they will achieve even higher peaks in the later 'honeymoon' stages.

Luckily, just as in ape consortships, the condition of being 'in love' is generally short-lived. The lovers come down from their temporary high and return to more normal styles of work and living. Often human courtship may lead to the formation of a longer-lasting social pair-bond, similar to those seen in gibbons and gorillas and to a lesser extent among bonobos and Sumatran orang-utans. The pair-bonds are frequently consecrated in some binding legal form of marriage and the real purpose of the pairing becomes blurred in social codes and rules. In evolutionary terms, however, these pair-bonds constitute the acceptance by the male of responsibility for, and assistance in the rearing of, any offspring that may result from his exclusive sexual relationship with his mate.

Why should the male accept this responsibility? Why does man not leave the responsibility and effort of child-rearing entirely to the female as do chimpanzees and Bornean orang-utans? He does not do so because his children would have a far smaller chance of survival if they were reared by the female alone. The male's evolutionary success, measured in terms of viable offspring, has been greater where he has put the extra effort into child-rearing and supporting and protecting the family unit.

Chimps and orang-utans have few predators and females are able to cope with the extra load of pregnancy, carrying and suckling an infant on their own. Gibbon and gorilla females are less capable and it is therefore in the interest of both parents to co-operate in a social bond to increase the survival chances of their offspring. The human mother, too, needs her mate's co-operation in child-rearing. She bears larger babies than the other apes, they are helpless and dependent over a longer period; they greatly reduce her mobility, her ability to gather food or perform other work. She is not as self-sufficient as her chimp and orang cousins and needs extra assistance and support. This is why our species has evolved a long-lasting pair-bond. The pair-bond has a strong biological basis and is reinforced in many biological ways. Desmond Morris in *The Naked Ape* has described in colourful detail how the female human became attractive, sexy and alluring to hold her man. In our species sex has developed as a means of strengthening the pair-bond. Sex has become elaborated, more varied, prolonged and more mutually exciting and pleasurable than in any other apes. Females extended their period of sexual receptivity and both sexes greatly increased their sensory apparatus in the genital regions.

But why should we go to such lengths to strengthen a simple pair-bond? Gibbons and gorillas form far more stable partnerships without any such palaver. They do not need regular shared sexual pleasure to hold them together; in fact in both species sexual behaviour is infrequent and performed solely for procreation. If it was clearly in the male human's evolutionary interest to form a permanent and stable pair-bond we would not have become so dependent on sex. Our society would not be riddled with the agonies of love and sex, adultery, unfaithfulness, prostitution, divorce and separation, sexual jealousies and prejudices. Women would be women and men men, their roles clearly defined and separate, singly meaningless but mutually complementary. Men would have been nothing without their women and women nothing without their men; both sexes would have been of equal importance. We would have had no male chauvinism, no women's liberation movements. So what went wrong?

Amazing as it may seem, the trouble has arisen because women are too competent for their own good. They can if necessary raise offspring on their own. If our babies were not so large and difficult to rear women would have been totally self-sufficient, we would never have needed prolonged pair-bonds at all, we could have enjoyed the brief sexual encounters of typical ape consortships and left our womenfolk to raise our infants for us as do chimps and Bornean orangs. On the other hand, if women had been totally unable to cope on their own, we would have become a stable pair-bonding species and there would have been no evolutionary advantage in cheating on the system.

Sadly, there has always been a fair chance that a female fertilized and deserted would still succeed in rearing her young, and this has given great genetic encouragement for male philandering – a phenomenon that strict cultural and social codes have never been able to eradicate. Men have always fancied a little bit on the side; have always had extra-marital relationships; have always misbehaved when they went off on hunting, warring or business trips outside their own social community; have never fully accepted the permanence or sanctity of marriage to the exclusion of all other sexual relationships.

If, for the sake of argument, we assume that a female has twice the chances of rearing her young with male assistance than without, we can say that the value of the pair-bond is to double her overall reproductive success. The male's position is somewhat different. By accepting a family role a male doubles his reproductive chances with his mate, but he does not double his reproductive chances overall. He could have the same reproductive chances by simply courting two females at the same time; indeed this could well involve less reproductive effort than maintaining one full-time pair-bond. Frequently males have got the best of both worlds by forming a pair-bond but also cheating on their wives and enjoying extra, irresponsible relationships.

The serious consequence of this state of affairs is that the formation of a pair-bond is more in the female's interest than the male's, and it is this basic inequality that has led to much of the socio-sexual strife in our society. The male sex is larger, stronger and physically dominant. It is the men who have been able to call the social tune. They have grudgingly accepted females as permanent partners only on their own terms. The females have been required to make themselves indispensable subordinates to their male master, to take on the domestic workload, to relieve them of the more tedious jobs and to gratify their sexual desires. Evolution has not only selected those females who could accept a subordinate role but has favoured females who actually enjoyed such a role. By establishing polygamous pair-bonds powerful males have been able to justify even more clearly their 'You need me more than I need you' attitudes. Even today in countries where polygamy is permissible women enjoy a much more subordinate status than in monogamous societies, just as female gorillas are more dominated by silverback males than gibbon females are by their monogamous partners.

The war of the sexes has been no walk-over. In making herself so sexually interesting the female of our species has more than made up for what she lacks in size and strength. Female monkeys and chimps enjoy a temporary boost in their status only when they are sexually receptive, but a woman's sexual interest never wanes; she remains at all times alluring and is sexually receptive throughout most of the month.

Beautiful women are among the most powerful members of all human societies. Although the male sex usually manages to dominate affairs at the community level, within their own families womenfolk are not so subordinate and are well able to get their own way.

There are, moreover, great advantages in playing a supporting rather than a leading role. Women have been able to enjoy high security and equivalent standards of living without the stress and responsibility that the provident and competitive male has had to endure. It is the man who dies in battle, the man who suffers stress-induced illnesses and the man who can expect to die at an earlier age.

Although the human pair-bond seems to be more in the female's interest than the male's, this does not mean that all selection has been for wifely fidelity. Cuckoldry has always had its advantages. The formation of the pair-bond is, in evolutionary terms, distinct from the selection of a suitable breeding partner. The function of the pair-bond is simply to guarantee that the female has the support and security that she requires to rear her young. The ideal pair-bond partner is therefore the best provider. The ideal bedmate on the other hand (sexual competence aside) is the best genetic breeding partner available. While there is often a strong correlation between a male's ability to provide and his suitability as a breeding partner, this is not always so. Despite very strong cultural penalties imposed against adulterous wives, cuckoldry has persisted throughout history. It is undoubtedly in the female's evolutionary interest to keep her eye open for genetically more suitable breeding partners so long as she can be sure that her pair-bond partner will remain ignorant of her infidelity and accept her young as his own.

Compared to the philandering male, the unfaithful female runs higher risks for much smaller advantages. Unless her partner is actually infertile she gains only qualitative benefits in her offpsring as opposed to the quantitative pay-off of more offspring for unfaithful males. Moreover she runs the risk of savage penalties if she is caught and has more to lose than a male partner from any breakdown in her pair-bond. Selection pressures favouring infidelity have consequently been much weaker in the female sex than in the male and there is thus a strong biological basis to the two sexes' somewhat different attitudes towards sexual freedom. There is indeed some truth in the male philosophy that it is more natural for the husband to be unfaithful in marriage than the wife.

Courtship patterns outside marriage follow essentially typical patterns, except that they must be conducted in greater secrecy and at opportunistically snatched or manufactured moments. There are times, however, when courtship is greatly abbreviated. Men have always been great wanderers, either because they are nomads, or are far from home on warring parties, hunting or business trips. Outside their own social communities they sometimes meet attractive females with whom they are unable to enjoy full consortships. Rather than waste such reproductive opportunities, males may attempt to abbreviate or shortcut courtship. If the female concerned capitalizes professionally as a prostitute, wishes to cuckold a less exciting partner or is enjoying pre-bonding sexual experimentation, she may well agree to such a manoeuvre. Normally, however, it is in the female's interest to resist advances from strange males who are not free to form lasting pair-bonds. In such cases the male's only alternative *reproductive* strategy is rape.

Rape is as difficult for the behaviourist to define as it has been for the lawmaker, particularly since it is simply an extreme form of normal patterns of courtship. Rape is the forceful sexual entry of an unwilling female partner but where do we draw the line? Since courtship is the behaviour by which one partner persuades the other to mate, there is some element of female resistance implicit in all male-initiated courtship, otherwise the courtship would not be necessary. Moreover elements of male physical domination play a part in many a licit and mutually satisfactory seduction.

Wherever we draw the line, however, it is clear that rape is the most selfish and anti-social of all forms of courtship. The rapist accepts no responsibility for his actions, and consequences must be borne by the victim, her family and her social community. It is not surprising that societies outlaw rape so strongly and that in many cases the death penalty is applied to this crime. Yet, despite this universal condemnation of rape, the behaviour has survived.

Among living apes only the orang-utan male shares the male human's problem of frequently encountering strange females with whom he is unable to form full consortships and so it is probably no coincidence that he, too, is a regular rapist.

The human male will usually only rape unwilling females *outside* the circle of his normal social contacts and because of the low reproductive pay-off and high penal risks he only commits rape when he feels that the chances of detection or recrimination are very small. Rape is common behaviour for armies of occupation. Indeed, together with pillage, rape is still regarded as one of the traditional rewards of military success. We have to regard such behaviour by males as natural, as having made good reproductive sense during our long warrior history. This does not mean that all forms of rape are adaptive; the cases that draw banner headlines in our newspapers are often the perverse results of social or psychological abnormality or disturbance. Nor does the fact that rape behaviour has been adaptive in our past and is consequently part of our natural biological repertoire in any way help to justify individual cases of rape. Man is in conscious control of all he does and must be answerable for all his actions. A better understanding of the biological controls and evolutionary nature of rape behaviour may, however, help us to devise more suitable ways of dealing with rapists and helping the potential rapist to understand and control his own feelings.

Rape causes enormous social problems, for not only is it distressing and socially destructive for the victim, but the very existence of such behaviour affects every one of us and contributes further towards the antagonism that exists between the sexes. As a result the entire female sex treats the whole male sex with distrust. Mothers view their daughters' suitors with suspicion. Every strange male is a potential rapist. The male sex is no less affected. Men know the strength of their own sex drives, we are each haunted by the question: are we really in control of our urges? What would it take to make us rapists? Have we never been tempted by the evocative bustle of a female form hurrying by in some lonely place? What holds us back? Respect for the opposite sex or of individual freedom? Or is it merely our social conditioning and the fear of cultural reprisals? Usually it is the latter, and with every rape case we read of we must share our sex's guilt.

Some instances of rape by young male orang-utans seem to serve more of a social than reproductive function. The male boosts his own ego by dominating adult females before he dares to challenge the established male for status in the dominance hierarchy. Many human sexual conquests, including some rapes, also serve a social rather than reproductive function. Young men are wont to boast about their female conquests to impress their friends in much the same way as a hunter flaunts his trophies. In some cases female acquiescence may also be due to a need to boost her own ego to prove she still has what it takes to pull a man.

Prostitution is another example of social rather than reproductive sexual behaviour. Prostitution is claimed to be the most ancient of all professions, and if you regard it as the proffering of sex not for reproduction but as a means of getting what you want from the opposite gender, then prostitution is clearly far older than our own species. Consider the low-fertility, promiscuous sex offered by receptive female chimpanzees to community males as a way of gaining acceptance within their protected communal ranges. Consider the false sexual receptivity by which female orang-utans are able to secure consorts while they are still carrying vulnerable small infants; the extended false receptivity by which the human female has secured the male's submission to family pair-bondage. Are not all of these forms of prostitution? Not that the human male at any rate has been in any way an unwilling partner; for in his role as procurer, pimp and patron he is every bit as much to blame for man's, and woman's, preoccupation with the sins of the flesh.

CHAPTER 10

Man and his Community

Human families are not self-sufficient. They live within the context of larger social communities and must co-operate closely with the other members of their village with whom they have almost daily contact. They must also co-operate, though less closely, with wider social circles, with folk from neighbouring hamlets, all villages that belong to the same tribe, all tribes that share the same language, other nations of the same race and ultimately with the rest of their species. This is nothing new, indeed it is very ape. Even gibbon and siamang families are only meaningful within the context of a population from whom they receive daily feedback from sightings, border skirmishes and chorusing behaviour. These lesser apes are interacting with the rest of the population, locating suitable mates, sensing gaps in the territorial pattern. Among gorillas several families share very close relationships within the relatively stable group, and different groups react together and exchange females in ever-widening social relationships. Among chimpanzees, bonobos and orang-utans the families are more individualistic, but the ranges of many families overlap and social interaction, tolerance and some sort of co-operation are therefore imperative. The sharing of male responsibility and the co-operation in hunting and range defence by chimpanzees, as well as their regular social participation in grooming and feeding activities, give the chimpanzee the most complex and advanced form of community shown by living apes. It is this chimpanzee level of community that we should keep closest to mind when considering the development and structure of the social communities of our own species.

Traditionally the individual human's social community was easy to define as the families that lived in the same village. Within large towns and cities, however, the community is more scattered and less discrete, but remains a useful concept for describing the network of close relationships that each individual maintains with his neighbours, relatives and workmates.

All hunter-gatherer peoples who have survived to historical times have very similar tight communities and a high level of mutual co-operation within these communities. We must infer that this type of social organization has been strongly favoured for as long as our ancestors have occupied a hunter-gatherer niche, i.e., something like four million years. Among living apes the species that shows the most similar co-operative community organization is the chimpanzee, and it too leads a hunter-gatherer type of existence.

I have already argued that the ancestral forest apes did not co-operate with one another, they were essentially individualistic animals and therefore selfish. But the climatic conditions which moulded the shape of hominid evolution forced selfish apes into a degree of co-operation completely contrary to their previous social evolution. The hominid community was formed out of jealous, competitive and haughty individuals. The potential for conflict and disruptive aggression within the community was enormous. We might even describe nature's efforts to bind such selfish individualists into co-operative communities as desperate, for to bind us together at all she had to use every evolutionary trick in the book. In human groups we see not only all the ploys used by primates and the sort of binding behaviour found in aggregating fish, birds and plains game, adult play as found in marine mammals, food sharing like that of large carnivores, clan relationships like those seen among elephants and caste systems as found in social insects, but we have also invented our own unique bonding activities: sport, music, conversation, humour, and other social pastimes.

There were several obvious advantages from co-operative behaviour which must have greatly encouraged the ape-men to band together into groups rather than range solitarily or in small family units. In a world of large carnivores, including sabre-tooth tigers, the Gracile ape-man, a mere four feet in height, was a puny fighter. Like the group-living baboons and other social animals of the plains, his chances of avoiding predation would have been greatly increased by co-operating with other males. Not only were there more scouts on the look-out for danger, but the males could band together to fight off any threat.

Hominids were also forced to band together for protection against raiders from other communities. Although larger bands could always exterminate smaller groups if they were determined to do so, this was usually not worth their while since they were likely to suffer serious casualties themselves. Banding together in mutual support is still the most effective way that a number of men can impose their common demands on the rest of society, a lesson of which we are constantly reminded by military alliances and trade unions.

Once co-operation was extended to the hunting party, individuals soon found they enjoyed more success than when they had hunted alone. It is impossible to hunt as a group without sharing out the bag. Such food-sharing behaviour also permitted the early division of labour between womenfolk and their men. Whoever provided the food all were fed, and this left some of the group free for activities other than food-gathering. The community in its food sharing behaviour acted as a welfare state; if an individual was sick, he would not starve but would be supported by the rest of the group who would benefit similarly if they ever fell ill themselves. Such altruism was in no way selfless;[1] it was in each individual's interest to be helpful towards others so that he could benefit in his own turn. The occasional valuable returns from the co-operative system more than outweighed the regular cheap inputs of community effort.

Another form of reciprocal co-operation was the early job specialization that occurred in human communities. It was more efficient to share out the community tasks so that everyone did what he was best at. Craftsmen were encouraged to develop their diverse skills rather than everyone trying to be a Jack-of-all-trades. It is impossible to define the ideal cricketer or ideal rugger player because successful cricket and rugger teams have to combine a variety of different talents. So it was with early human communities. Men fought, worked and hunted as teams and there was always selection for a diversity of physical and mental faculties.

As food requirements or defensive needs varied, so the groupings within communities could expand or disband. Hunter-gatherer communities today still retain that ancestral ape characteristic of great social flux. In seasons when food is short the community might break up into small sub-units which forage separately; at other times two or more communities might join to share some abundant food supply or face some common challenge.

Co-operative hunting or fighting binds the participants closely together, especially if danger is involved. If you have to entrust your life to the watchfulness and skill of your comrades, you very soon establish strong ties and friendships with those comrades. Look at the comradeship of soldiers in hazardous war-time campaigns, or the closeness of mining and fishing communities. This comradeship is strengthened by repeatedly recounting the group's shared experiences, comparing each performer's part in the action.

Man's ancestors needed to band together in co-operative defence against fellow man. Actual attacks were probably very rare but when they did occur they could prove lethal. It paid hominid bands to practise their co-operative tactics and it was also vital that the different males involved could all enjoy high, relatively equal status, i.e., that there should be many chiefs and few Indians. The solution to both problems proved to be regular participation in social hunting. Hunting originated not as a means of getting protein into the diet but as a substitute to fighting fellow comrades.[2]

Traditionally male primates achieve high status by demonstrating their physical prowess in social fighting. Fighting between males of the same co-operative community, however, would not only disrupt and weaken the alliance, but might result in serious injury. Only by redirecting aggression away from their own community, by raids on neighbouring communities or in hunting, could males achieve high status without reducing the effectiveness of their own community.

The origins of hunting large vertebrate prey by baboons and by chimpanzees are probably identical. Like early man both these primates have also faced ecological pressures that have forced many adult males to join together in a co-operative monkey group and ape community respectively. They are almost unique in the levels of inter-male tolerance and co-operation they exhibit in both their everyday interactions and also occasional fights against neighbouring groups or threatening predators. As in human groups the hunting of large vertebrates is an almost totally male affair and the protein gained in such hunts could have as easily been obtained by gathering invertebrates.

Gorillas and orang-utans do not hunt large prey because, although they probably require just as much protein in their diet, they have no social need for hunting. Male tolerance and co-operation are low; male orang-utans are solitary and most gorilla groups have only one silverback patriarch.

With hunting as a substitute for social fighting, the most effective prey were those that most closely resembled the hunter's own species. The preferred prey of chimpanzees are other primates, and human hunters also hunt monkeys and apes wherever these are available. In Borneo the Iban Dyak head-hunters hunt orang-utans as human substitutes now that they are discouraged from taking their gory trophies from the Kayans and Chinese. Moreover since great respect was always paid to the successful hunter who made the kill, hunting provided opportunities for many males in the same community to achieve high respect and status. By hunting as a team, all the hunters in the party got a share in the credit for a successful kill, as well as an opportunity to practise their fighting tactics.

Men and apes are poorly equipped for digesting raw fresh meat and are sick if they eat large quantities. It is probable that meat only became a major contribution to the overall diet when man had discovered cooking – long after his ancestors were fully bipedal and long after they were social hunters. Protein never ousted carbohydrates as our species' staple diet. Even in modern industrialized communities we can still see this pattern of hunting for glory rather than food. Why else should hunters keep animals' heads as trophies of their prowess and hunt animals they have no intention of eating?

We can view the development of competitive team games in precisely the same way, as another play form of battle, during which the participants learn to co-operate with one another. Team games have developed on all continents along amazingly similar lines. Despite the competition for team places and against a background of possible rivalry for community status, females, property, etc., once the game starts the team members must forget their differences and co-operate with one another if they are to have any hope of success. Prior to the start of the game, whether it be rugger, soccer, Indian lacrosse or Papuan cricket, the players must prepare themselves psychologically, chanting, praying, discussing tactics, warming up the muscles, preparing the mind and body for the forthcoming trial. After the game has ended the discussion begins, the teams retire to the communal 'baths' or food and drink to re-run the great moments of triumph and failure; the pseudo-battle is analysed in minute detail, the glory and kudos are allocated. The team becomes symbolic of the social group from which it is drawn and has ranks of ardent supporters who are also brought closer together through their shared support of the team.

Communities are held together by a whole host of other activities. In tropical countries it is the custom at the end of the day to join in the phenomenon of evening assembly. The day's work is over, now is the time to relax, to change into fresh clothes and come out of houses and huts to meet, to show off, to enjoy social encounter and conversation. Participants chatter about unimportant affairs, local gossip or even business matters. The old folk recall the good old days and wonder what the world is coming to. The young girls promenade coyly in twos and threes; small gangs of youths eye them up, making suggestive whistles and comments. Young men woo their girlfriends, mothers scream at their naughty children, men escape from their wives to chat with a neighbour or workmate. The whole of the human social condition is on display but behind the deluge of tiny individual activities, and largely unrecognized by the performers, everyone is participating in a communal event.

As dusk approached I walked along the road towards a town in central Zaire. From a mile away I could hear the buzz of humanity. The occasional piercing shout or laughter broke through the tumult and suddenly I became acutely conscious of the function of the clamour. I felt intimidated by the very foreignness of the sound. That was not my community nor even my race. To me, an alien, a stranger, the noise betokened a powerful and potentially hostile group to which I did not and could not belong. To the individual participants, however, the busy hum of the village served the reverse function. This was the noise of their own community, here they were accepted, surrounded by comrades and security. The clamour meant that all was well, the village was in good spirits, everything was fine. Although most of the promenaders would not recognize this message of well-being in the evening's chatter, they would certainly know that something was wrong if there was no noise, if the village was silent or filled with the pitiful wails of mourners bemoaning the recent death of a loved member.

Participation in such community-binding group noise does not seem so different from the territorial message of the gibbon family chorus – a sort of 'We are well, we are strong, we live here; if you're with us rejoice in our strength, if you're agin us beware'.

The same message is inherent in the contact calls made irregularly through the day between different foraging parties of chimpanzees that belong to the same community. At night when the chimps are tucked safely in their nests they again make contact calls letting the other scattered nesters know where they are. More than location information is involved; it is the participation in the sum group sound that helps the individual strengthen his community membership. Not only is the function of the chimp chorusing very similar to the human behaviour of evening assembly but the very sounds are almost identical. Primatologist Vernon Reynolds[3] plays a lecture tape to his students of chimpanzee group chorus vocalizations which imperceptibly merges into a tape of human group chorusing in a cocktail party. Individual meaning of language components are lost in the hubbub of grunts and shouts, shrieks and giggles. Only when a give-away peal of laughter breaks through the mêlée of other sounds does the audience realize it has been listening to its own species.

In colder climates the evening assembly is not a practical reality but we meet indoors instead, in pubs, clubs, at parties. All these fulfil a similar social function, and the Sunday church service brings the whole community together for shared participation. Even such modern substitutes as television and newspapers fulfil a similar need. We read the same columns and follow the same series, watch the same newscaster reporting on the latest happenings in the larger society to which we belong. We keep in touch with our own kind, thus reinforcing our own group identities.

From time to time the community makes an even greater effort in the form of an organized occasion, a feast, festival, carnival, show, concert, display or public holiday, but whatever the form the function is the same. The community is strengthened, and more closely linked, by its members sharing pleasurable activities together.

Human communities are further strengthened by complex webs of family relationships and inter-marriage, adult and child play relationships, relationships between workmates, reciprocation between neighbours, the borrowed cup of sugar and shared pot of tea, professional and business relationships, not to mention the links of language, culture, race and religion binding neighbouring communities together in ever-widening social circles. Within the community there also develop tight co-operative agreements in the form of small, exclusive clubs and secret societies. Many of these institutions have initiation rites which wrap a veil of mystery and superstition around their membership and acceptance. Pain renders the initiation rites more memorable, more permanent. African, South American and Papuan tribespeople scar and mutilate themselves or pierce their lips, noses and ears with bone and wooden plugs. Dyaks and Maoris tattoo themselves. Australian aborigine war-parties mingle the blood from their freshly mutilated penises. Both sexes of Kikuyus undergo painful circumcision when they reach puberty. The initiation rites of boarding schools and street gangs may be less gruesome but are functionally identical. In all cases these societies or clubs have some signal by which members can recognize one another, whether this be a tattoo, a tie, a password or a freemason's hand-clasp.

Our inherent craving for the types of social interactions that form such co-operative relationships shows how important these relationships have been through our evolutionary history. The male orang-utan not only does not miss the company of other orangs but resents their proximity when they are close, but humans are encouraged to congregate by their need for acceptance, their love of social intercourse and their hatred of loneliness. Solitary confinement has long been recognized as one of the cruellest ways of torturing a man. Even removed from kith and kin, workmates and comrades we seek out social contact, we start to chat, to talk about the weather and, in less inhibited societies, even groom one another. Even in western societies we will allow total strangers to groom us providing this occurs in a proper social context. We feel quite at ease in the hands of hairdressers, masseurs and manicurists, even prostitutes, and it is often with just these strangers that we feel most able to discuss our most vulnerable and embarrassing social problems.

In crowd situations people have a remarkable and still poorly-understood capacity to think in accord. Whole crowds of strangers will 'ooh' and 'aah' with infectious sympathy as they watch a football game or break into spontaneous and individually uncharacteristic noisy applause or hostile behaviour because the individual's inhibitions have been submerged by the strength of the feelings of the crowd. Audiences weep and laugh together, wax violent, repentant or calm together. Just like a shoal of fish turning in complete synchrony we are swayed by the emotions of the crowd around us and it takes a conscious and determined effort not to follow the will of the mass. I remember attending the funeral of an old Bornean lady whom I had never met and being so moved by the chanting crowd that I wept with the other mourners as the coffin was carried to its final resting place in a forest cave.

We are similarly in tune with the crowds of strangers we meet in a busy street and can recognize at a glance the intent of those approaching us, who will give way and who will not. Without conscious thought we move through a mêlée of people all going in different directions at different speeds with rarely a collision. We instantly recognize and give way to the purposeful stride of the man who will not deviate and just as quickly recognize the tiny signals that indicate that another stranger is willing to step aside. Try to walk in a straight line down a crowded pavement and you will probably fail, defeated by a pram-pushing young father or middle-aged owner of a pink-coated poodle. I have tried this apparently simple manoeuvre but my appearance obviously belies my intent, the system breaks down and I collide hopelessly. Just occasionally the signals are misread, two pedestrians move aside in the same direction, then step hurriedly back and are caught in a comical impasse until one or other finally succeeds in passing.

The behaviours I have just described are extraordinary for a primate. They are the sort of crowd behaviours only otherwise seen in creatures which aggregate in large numbers, such as shoals of fish, flocks of birds and the herding ungulates. Small bands of hunter-gatherers can hardly have needed such behaviours and man has lived in crowded urban situations for only a tiny and recent part of his history, too short a time to have developed such social awareness from scratch. It is far more likely that the close co-operation needed during the ancestral hunt encouraged the hunters' uncanny telepathic ability to read the intentions of their fellows and this ability has now proved to be invaluable in the crowded situations in which we find ourselves today.

For all his attempts to maintain and strengthen the community, man is still a jealous, greedy, competitive, selfish ape, and there is great rivalry and stress both within and between neighbouring human communities. For the individual the community provides food, security, friends, potential mates and somewhere to turn in times of trouble. On the other hand, the community comprises a threat in the form of many rivals with whom to share food, land and other property and especially the best breeding partners. Herein lie the roots of so much human aggression.

Human aggression can be narrowly defined as those activities motivated by the intention to hurt or injure members of our own species. At the other extreme, human aggression can be interpreted as man's forceful rather than intellectual attempt to improve unsatisfactory elements of his environment. While it is the human violence aspect of aggression that constitutes social problems, I do not think that this aspect of aggressive behaviour can be usefully divorced from the wider definition. For there rest many of the rules that govern our aggressive expression and the adaptability of aggressive behaviour in our species.

Before discussing the nature of human aggression, I would like to dispose of some rather dubious ideas that have greatly clouded the issue. Firstly it is frequently asserted that human aggression is in some way different from that in animals, that everything animals do is natural and therefore pure, whilst human aggression is evil because only humans commit murder, wage wars, enslave their defeated opponents, kill for pleasure, perform infanticide, genocide, suicide and cannibalism. These ideas are totally false, based on ignorance and rather masochistic intellectual indulgence.

When a male grey langur or a male lion takes over a group by ousting the old patriarch he proceeds to kill all the young in the group. The mothers are powerless to prevent him and it is clearly in the new patriarch's interest to perform this slaughter since it results in the adult females coming back into reproductive condition quickly, giving the new male a better chance of leaving his own young. A dominant female hunting dog murders the pups of other bitches, pups that threaten the survival of her own.

Immediately after emergence, queen bees kill the inhabitants of their sister queen cells. Guard bees at the entrance to a hive sting and kill any strangers that try to enter, including their own siblings if these have been removed from the hive as pupae and attempt to return as newly-emerged adults. They smell wrong and are killed. Herring gulls eat the eggs and chicks of their neighbours if these are left unguarded. Rooks steal sticks from the other building colonists with such regularity that every pair takes several times as long to make its nest each spring as it would if there was no pilfering.

Male lions and deer stags are sometimes killed or seriously injured during their battles over mating rights. Hyenas may kill and even eat other hyenas. During chimpanzee border skirmishes combatants may be injured and innocent young killed and even eaten. Ant armies wage war on one another; they make slaves of lower caste members of their own and other species. Old trout have no qualms about eating smaller trout; spiders and mantids often devour their mates after or even during copulation. Many carnivores show obvious pleasure in the kill, some will kill for killing's sake far in excess of what they can eat. Let a fox loose in a hen run and you will see a picture of wanton destruction. Most rodents will start on their comrades if there is nothing else to eat. As we delve deeper into the behaviour of our fellow animals we find ever-lengthening lists of such 'atrocious' behaviour. We are forced to conclude that in nature there are no evils. Anything goes so long as it is in the evolutionary interests of the performers and so long as the benefits from the actions outweigh the costs and risks involved.

In most cases the benefits do not outweigh the risks and animals settle their disputes by means of ritualized combat, tests of strength or predetermined dominance hierarchies. Exactly the same rules apply in man, who also settles most of his competitive conflicts without serious violence. The only difference between man and animals is that man is a fragile creature, easily killed, normally unarmed and often off-guard, whilst man's weapons are devastatingly effective. A wolf could never be sure of victory even if his rival was asleep when he attacked. Man's opportunities for murder are far greater; the skirmishes between neighbouring chimp communities have evolved into human tribal warfare.

Ardrey[4] and Lorenz[5] argue that man is essentially an aggressive territorial animal. Not so, say the social anthropologists[6]: he is basically very tolerant of his neighbours and enjoys a fluid exchange of members between different communities. The truth, of course, lies somewhere in between. Semantic generalizations are completely inadequate to describe a complex subject like human spatial arrangements. If by territorial we mean that men will defend land within certain boundaries for their own exclusive use then some men do and some do not. There are countless examples of warring town kingdoms, clan and tribal boundaries that fall pretty well into the strictest category of group territory. The warlike Dugum Dani of New Guinea actually survey their clan boundaries from rock outposts or specially-built wooden towers which they ascend each morning to yell abuse against their neighbours, just as gibbons do. In addition to such communal territoriality we also see rather aggressive possessiveness applied to land and even personal space as well as to breeding partners, material belongings, and money that could be described as individual territoriality.

Social anthropologists prefer to look to the few remaining hunter-gatherer peoples to answer problems of the basic 'natural' behaviour of man. The first descriptions of the Australian aborigines described these people as territorial and warlike; more recent studies have claimed the reverse. In fact the customs and practices of these peoples vary from tribe to tribe and have been so disturbed anyway that it is difficult to reconstruct their original patterns. Many tribes have been displaced so that their tribal rocks are now outside their own hunting grounds and those who wish to find their roots or commune with the ancient spirits must stray on long 'walkabouts'.

Certainly the African pygmies, Tanzanian Hadza, Kalahari Bushmen, northern Eskimoes, Malayan Semang and Philippine Tasadays show little overt territorialism, but there are good reasons why these survivors may not reflect the original patterns of hunter-gatherer societies either. The pre-agricultural peoples who have survived are those that constituted little or no threat or competition to the culturally dominant agricultural tribes who displaced them from all their best habitats. Hunter-gatherers survive in areas which are agriculturally useless and as long as they are unaggressive neighbours they are eminently tolerable and often useful trading partners. Most aggressively territorial hunter-gatherers have probably been exterminated and those hunter-gatherers that live in very severe climates, desert, dense forests or tundra no longer fit our image of man being his own worst enemy. In these peripheral habitats the environment is the real challenge, and the hunter-gatherers have enough to do merely to survive without wasting undue effort in fighting one another as well. Similarly the siamang cannot afford to perform his territorial display as often as the smaller gibbon because he has to spend more time and effort on finding his daily food.

Social anthropologists have undoubtedly over-reacted against the territorial aggressive image of our ancestors by describing their subjects as unrealistically unaggressive, but there is plenty of evidence that even the most peaceful hunter-gatherers do have deep-rooted aggressive and territorial attitudes. Homicide, especially in rows over women, is almost as frequent as in our most violent urban societies and territorial patterns are evident in their folklore and history. Much of their culture is designed to avoid aggression. Bushmen and pygmy bands will split rather than suffer the stress of unresolved conflict between different parties. Entering someone else's land is only permitted within well-established rules of conduct by the tolerated intruder. Food-sharing patterns are not so selfless as they are often made out to be; if a hunter kills a small animal it will get no further than his own family, if he brings it home at all and does not eat it himself in the bush. Larger game might be distributed to a few neighbouring or related families but only a very big piece of meat will be distributed throughout the whole community.

The fashionable Utopian picture of hunter-gatherer ways and attitudes seems to me as distorted as the earlier picture of primitive violence. Overall, I believe the surviving hunter-gatherers are no more or less selfish, no more or less violent than modern agriculturalists, urbanists or industrialists. What impresses me most is the similarities of behaviour shown by all mankind, how very alike we all are in spite of differences in race, religion, wealth or life-style.

Human aggression begins at birth, for the new baby soon discovers that by crying he can bully his mother into giving him the sort of environment he demands. Crying, pestering, temper tantrums and other forms of mother-bullying continue throughout infancy and sometimes right through life. Aggressive behaviour towards play-peers is important in determining status relationships, who is leader of the group, who gets first choice of the toys and first turn with the cricket bat.

Temper, tantrums and physical violence appear in early human childhood in frustration situations, just as they do in young apes. If all else fails, try using force. If force regularly brings rewards the individual's aggressive tendencies are bound to be encouraged. Normally, however, the child soon learns that force is only effective some of the time. It is no good trying to move the immovable, it is dangerous to use force on people or animals more powerful than oneself. A reputation for aggressiveness will make you a social outcast, the child that nobody wants as a playmate. Each individual learns early in life how to control his aggressive tendencies, to find peaceful solutions to difficult problems; he learns patience and tolerance, how to cope with frustration either by bottling up his emotions or redirecting his aggressive drives in less harmful, more constructive directions. Wherever it has paid to be aggressive, however, aggression has flourished and continues to flourish. We are all equipped with the body, glands and brain that allow us to develop whatever levels of aggression are most suitable to our circumstances, but we also have high levels of perceptiveness so that we can evaluate how appropriate aggressive behaviour would be in any situation, and we have sufficient conscious control of our own behaviour to act according to our evaluation.

Cultural pressures have applied social codes on the use and repression of aggression, but these codes have always made a clear distinction between behaviour within the community and behaviour towards outsiders. The ten Biblical commandments are absolutely typical of many tribal rules, but most of the commandments can be legitimately broken outside the tribe. The Israelites clearly did not see their new commandments as imposing any restrictions on their behaviour towards the Canaanites whose land they were about to capture through violence and deceit. In many cultures the rules are even more explicit. The Kikuyus were taught to believe that thou shalt not steal, except from an enemy, thou shalt not kill, except thine enemy, etc.[7] Clearly there has been selection favouring appropriate levels of caution but there has never been strong cultural or genetic pressure to check man's aggressive impulses completely and, as we know, warfare has been very much a way of life for much of our history.

Within communities there has always been a strong taboo against murder, and breaking this taboo has always incurred the highest penalty. All physical violence is discouraged and the settlement of individual disputes over sexual jealousies, ownership or inheritance is largely predetermined according to rank, size, age, wealth, class and order of birth, but failing this we have vocal and legal channels of argument to which we can resort. Nevertheless the right to resort to violence if all other attempts at settlement fail has always been recognized and many communities have codes or rules by which such fights might take place fairly and without great danger to either the participants or the stability of the community. Numerous forms of ritualized fighting, duelling, boxing, wrestling, arm levering, to name but a few, have arisen from such codes of conduct. They have arisen for the same reasons and in the same way as the tooth taboo of fighting chimpanzees within the same group.

In many cases the interests of the individual and the social group to which he belongs are the same and many social taboos probably have strong biological bases. There are, however, many instances in which the individual's interests are not the same as those of his community and herein lies the source of most community violence. The community can only penalize the selfish code-breaking behaviour that it detects. If you are not caught, cheating the system has consistently paid off and has therefore never been eradicated from our species' behavioural repertoire. Whilst it has been in almost every individual's interests to preach and pay lip service to the community codes and taboos, it has also been to his benefit to cheat whenever the stakes have been high enough and the risks of detection low. Sadly this is still the case. The majority of thefts and murders still go unsolved; selfish but intelligent crime continues to pay.

We do not have to delve far back into our carnivorous or cannibalistic past to find reasons why man kills his fellow man;[4] the reasons are clearly evident in every large human aggregation today. From the incidence of war and the ease with which decent friendly folk can be persuaded to take up arms against complete strangers who are painted as hostile, it is clear that we have no biological taboo about killing our own species. We may, however, have some biological as well as cultural taboos about killing individuals we know personally, but these finer feelings clearly do not extend to strangers. Unhappily, when sufficiently tempted or threatened, man also shows his willingness to kill his closest kith and kin.

We are right to be concerned about these socially destructive and undesirable traits in our species' make-up. We are right to try to understand why we show aggressive behaviour, why it was acquired and how it is controlled. We are right to try to reduce the incidence of such anti-social violence. But we are wrong to conclude that our anti-social behaviour gives any indication of something basically amiss, evil or in any way less natural than the violence shown in the rest of the animal kingdom. In fact the human species can claim to be among the most tolerant of all social species. The actual occurrence of serious physical violence among adults is fairly rare, far less frequent in fact than in any other social primate. The picture we have of a consistently high level of human aggression is a direct result of the crowded communities in which we live and the excellence of our world-wide communication systems through which we learn of wars, murders, riots and fights. If we had comparable knowledge of animal species we would realize what remarkably peaceful, tolerant and co-operative creatures most of us really are. When our observations of wild apes move out of the thousands of hours bracket towards the millions of hours bracket we shall see the rare instances of murder, cannibalism and other 'atrocious' behaviour in clearer proportions.

CHAPTER 11

Apes of the Future

We have traced the evolutionary histories of man and apes and seen how their failure to defend the simian forest niches against the socially-advanced monkeys forced our forebears to become large, tail-less, brainy, arboreal fruit-eaters. We have seen how climatic and geological pressures trapped one branch of forest apes in the savannah woodlands, forcing them to co-operate for defence and adopt a diet of grass seeds and root tubers. With this new way of life our ancestors became bipedal, made tools and indulged in social hunting.

Environmental instabilities of the last three million years, together with the constant threat of warring neighbours, have encouraged the trend of increasing intelligence and cultural competence, far greater than anything seen in the other apes. Man's behaviour and community structure remain essentially ape-like however, and our cultural practices have developed in a logical and rather predictable progression from foraging ape – hunter-gatherer – natural harvester – agriculturalist – urbanite and ultimately, within the last few centuries, to modern industrialist.

Can we now extrapolate from the path we have followed to look into the future? Can a better knowledge of ape ways and ape societies help us plan our own destiny as we feel our way into the unknown? Our technology is way beyond anything we have previously experienced. We find ourselves in the largest and most complicated social system of all animals. We have almost succeeded in eliminating disease and starvation as natural checks on our numbers. We have taken our destiny out of nature's tender hands and hold it so precariously in our own that we could easily annihilate ourselves within the century. We are facing problems that have never been faced before. Natural selection will be of little help. From here on we must rely on our own group intelligence. What sort of future can we expect? Indeed, do we have a future at all or will we fade into obscurity like the extinct dinosaurs, whilst the world waits for a new super-ape to evolve?

First let us consider the geological uncertainties upon which our survival depends. The sun of our solar system is good for another five billion years before its hydrogen becomes exhausted and it swells to a red giant and roasts all life on our planet. There does not seem much cause for immediate alarm there. The amount of radiation from the sun fluctuates slightly but there is no reason why we should expect world-wide climatic shifts any greater than those we have already survived. The Ice Ages of the Pleistocene were as dramatic as any event in the geological record, but a tremendous wealth of plant and animal forms survived intact and man emerged from the experience greatly advanced.

The continental shelves on which we live are constantly moving over the mantle of our planet. Continents will continue to tear apart like the rifting of Arabia from Africa and the rending of California from the United States. Such movements will cause earthquakes, tidal waves and volcanic eruptions but these will only be local disasters, no serious threat to the species as a whole.

We should be more worried about the man-induced changes in our weather patterns, vegetation and climate. Deforestation for timber or merely clearing bush for agriculture and living space has drastically affected the weather in many parts of the tropics. These tropical soils are poor and can support luxuriant forest only because in their millions of years of history the tree community has evolved together to protect the soil humus and nutrients from erosion by sun, wind and rain. Nutrients are held in living tissue and whenever a tree dies they are speedily broken down and released for re-use in the cycle of life. After such jungle has been cleared or burned the soil may support a few good crops but then the goodness is gone, it deteriorates to acid grassland fit only for a few cattle or goats. Grazing stock in turn destroy the habitat, eating everything in sight, and when the eroded land can no longer support even goats nothing can live there. Such deterioration is frightening enough in itself but the effects are more far-reaching than this. The large areas of bare land cause heat thermals that drive many rainclouds away and when rain does come there is no plant sponge to soak it up. The rain simply runs off in a flood, leaching away yet more of the soil.

The area of desert on our planet increases frighteningly every year. It has done so since man first took to agriculture and animal husbandry. Poor land use in India has transformed a lush, forested subcontinent into a land of thorny scrub with only eighteen per cent of the land still covered in natural vegetation; yet only six per cent of the cleared ground is still fit for agriculture. The Indian population suffer the consequences of three thousand years of land misuse, but elsewhere these changes are happening far more rapidly. Every year newspaper headlines scream of drought and flood disasters; so many of them are man-made and the position is not likely to improve. We have transformed the weather patterns almost irrevocably in many places. The tracts of green vegetation, which fix carbon dioxide from the atmosphere and release oxygen in its place, contract each year and, as the forests shrink, the level of oxygen in the atmosphere falls too. This level is only twenty-one per cent to start with and we would be in real trouble if it were reduced a few more per cent.

In short-term economics it may pay us to rape resources and to hell with the long-term consequences. But for every pound a man gets for chopping down a tree the world becomes a little cheaper for everyone else. The individual's best interests frequently conflict with his community's, the community's best interests may conflict with those of the nation and the nation's with the rest of the world. Even if we had the wisdom to see the right path we do not have the authority or the persuasiveness to make everyone follow it.

There is a timely lesson to be learned in the jungles of Central America. There, in 1839, the archaeologist John Lloyd Stevens discovered the abandoned temples, statues and cities of a great tropical civilization – the empire of the ancient Maya. Where the jungle now rules unchecked there once stood a dozen large cities housing a huge population of sophisticated people, a warrior race who had invented hieroglyphic writing and whose calendar was as accurate as our own, who could predict the solar and lunar eclipses and understood the movements of the stars and planets. What dreadful fate could have befallen so great a people that they abandoned their cities to the jungle? The Maya were destroyed by their own ecological inefficiency. The intellectual aristocracy were too engrossed in their erudite astrology to pay much attention to the agricultural practices of the peasants who provided for them. They knew nothing of the plough nor the wheel and continued to employ destructive shifting agricultural methods until high densities forced permanent cultivation of soils too poor to support such ravages. Rather than adopt a policy of proper land management they turned to human sacrifice as a cure for falling soil fertility. But the Maya civilization had expanded inwards from its geographical periphery and the time came when all the land had been cleared of jungle, planted and eventually impoverished. Now the peasants could no longer afford to pay the crippling fifty per cent taxes to the ruling classes and the whole socio-economic system collapsed. After a century of peasant uprisings and mass migration the entire civilization was deserted. The ruling classes, with Toltec aid, began to build a new empire in virgin jungle two hundred and fifty miles further north, but they never achieved their former strength and were unable to put up any resistance to the Spanish Conquistadors six hundred years later.

We should be wary not to fall into the same trap as the ancient Maya. We should heed the ecologists who preach doom and disaster. It requires very little of a nation's budget to monitor the impact on the environment of its industry and agricultural practices, yet few countries have so far shown any responsibility in this direction. We recklessly continue to rape our forests, soils and limited reserves of phosphates and fossil fuels. We pollute our land, our atmosphere, our waterways and even our own bodies. Every year the earth becomes a poorer place to live in and our wasteful selfishness piles up further problems for future generations to face.

Apart from protecting our physical environment we must also tackle many growing social problems. As he has expanded in numbers man has adopted an increasingly complex social order, and over the centuries we have tended to form larger and larger military alliances. We are now in the frightening position of one half of the world lined up against the other half and of both sides escalating their build-up of sophisticated weaponry. Should these weapons ever be used they are sufficiently powerful to completely destroy civilization, if not humanity, on our planet. There is some security in the fact that neither side could achieve a nuclear victory without suffering almost total destruction itself, but will that always be true? Will one opponent devise a perfect nuclear shield? Will one side be betrayed by those in charge of its defences or will we find ourselves subjected, not to atomic power, but to an onslaught of germs, poisons or sterilants? History can give us no comfort, for there have been no other occasions where such weaponry stockpiling did not result in war.

The logical next step should be for the whole species to unite into one single military alliance, but against whom? We are scanning the universe for signs of other intelligent life. There is no reason why there should not be such intelligence, no reason why in fact there should not be very human-like creatures somewhere among the billions of other planets in our own and other galaxies. Even if we do not make contact we could invent an external threat. As a species we have always needed common threats and challenges to bind us together. In the absence of real dangers we have often had to invent threats in the form of ghosts, monsters, evil spirits and devils. We have pictured many of these mythical fears in the image of super-apes, so what is more logical than to now convince ourselves that somewhere in the star-filled universe live hostile supermen and that we should prepare our united military defences for a possible invasion?

There are other more realistic and immediate threats which could be tackled on a united front. We must draw a halt to over-exploitation of the world's limited resources of minerals, fossil fuels, marine fish and mammal stocks and natural forest and to our disgusting habit of polluting our own planet. These are problems that must be faced at international level if we are to find a solution. So long as even one nation refuses to fall into line all could be doomed.

Two thousand years ago Christ preached that if we are to be saved we must extend our concept of neighbour to embrace the whole of humanity. It was a novel idea at the time but the truth of his teaching has never been clearer than it is today. If we are to combine into one huge co-operative community we must follow well-established behaviour patterns of increased socialization. We must try to erode national boundaries, take up a shared language and enjoy free cultural exchange, sharing our scientific and technological knowledge and our medical advances. We must integrate the races more freely by regular, friendly co-operative intercourse. This does not mean we should indulge in mass interbreeding to blur genetic barriers, for this would be foolishly wasteful. Different racial gene pools are closely adapted to suit local environmental conditions and we should exploit these specializations rather than destroy them. There are differences between the races and it is foolish to deny them, but it is as futile to claim that one race is superior to another or that one individual is superior to another as it would be to claim that a dog is superior to a cat. A dog may perform better in some ways but it is certainly inferior in others. Every other individual can do at least something better than we can ourselves and in certain circumstances that talent could be the difference between their surviving and our not. How can one compare the value of intelligence against strength, skill against speed when, according to the circumstances, any one of these qualities could have the supreme value of saving the individual's life? Each race has its own particular contribution to make, just as players of different physique and abilities are needed in a rugger team to improve the performance of the team as a whole. We should not try to hide or overlook racial differences, whether these be of size, shape, colour, temperament, athletic, artistic or intellectual leanings; rather we should recognize the different adaptations and make the most of them. Some racial interbreeding may well prove important in creating a wealth of new genetic variability on which selection can continue to work for the improvement of the species, just as cross-breeding has always been the main tool in improving our domestic stocks of animals and plants.

Even competitive encounter can be socially bonding so long as it is peaceful. Sport is a regular example of this and no one who visits the Olympic villages can fail to recognize the tremendous bonds that link the athletes of many nations, although it is their personal and national competitiveness and divisiveness that has brought them together in the first place.

It is natural enough to learn hatred for those people with whom you have repeated hostile interactions, but it is not so easy to hate complete strangers unless you can invest them with characters that threaten all you hold dear. Once you meet the 'enemy' and realize that, however foreign, he has likes, dislikes and attitudes similar to your own then it is very difficult to hate or kill him simply because he is foreign.

I think it is for just this reason that we have seen over the last few decades a voluntary abstention and progressive withdrawal from active warfare by the Western Caucasian races. At the end of the Second World War the Caucasian races dominated this planet numerically, economically and technologically. They had a complete stranglehold on world power structure, but they lacked the motivation to take all that was free for the taking. The European countries gave up their colonies not because they could not hold them against the local nationalist movements but because they were unwilling to retain them by force from people who no longer seemed very alien, no longer seemed a threat, no longer fitted the concept of an enemy.

The Caucasians have travelled more than other races, mixed with many nationalities, they have learned about foreign ways from books, films and television. They have laughed with, loved with and cried for other races, assisted, educated and converted them. They can only fight other races if they feel genuinely threatened. In the most recent and dramatic example, we have seen how America, the most powerful nation on earth, was unable to win a military victory over a patriotic crowd of poorly-equipped jungle guerrillas because the young Americans could not put their heart into killing people who were clearly no direct threat to America and were no more foreign than their South Vietnamese allies and bedmates.

War has lost a lot of its old appeal. There was fun and excitement in the days of the axe and sword, even of arrows, rifle bullets or aerial dogfights. Man loves the challenge of face-to-face competition, the glamour and glory open for the valiant to grasp. Men could display their war medals as proudly as the Iban shows off his head trophies or the Boran tribesman the testicles of his enemies. There is no glory left in today's contests, however, for it is heartless technology that is put to the test, not men's strength and spirit. I am sure that if other countries could break down their isolationism, as the Caucasian races have done, and would participate more freely and peaceably with the rest of the world, the incidence of large-scale ethnic wars could be cut enormously.

I think that a better understanding of the rules that govern the socialization processes of apes and other animals could also help to reduce the outbreak of small-scale wars that do result from genuine competitive threat and reciprocal atrocity. At present such protagonists tend to seek the military support of larger, wealthier nations, and although these latter may imagine they are preserving peace by maintaining a power balance they are in fact escalating both the capacity for an outbreak of violence and its likelihood. A far more effective way of keeping conflict under control would be the threat of deprivation. We have recently seen the political effectiveness of an Arab oil embargo. If several of the world's major commodities came under international control there would be few countries sufficiently self-sufficient to risk breaking the international codes of peaceful settlement for disputes. The attempted embargo of Rhodesia failed to bring down the UDI, not because the method was weak, but because so many countries were better off ignoring the embargo and maintaining their

Rhodesian import and export markets.

Mutual interdependence is the force that drives groups of apes, people or any social creatures together into co-operative systems, and if countries are to live in peaceful co-operation the same principles must be put into effect. Iceland could force British fishing fleets to leave Icelandic waters because she was not dependent on Britain to any reciprocal extent. The same lack of interdependence characterizes all the world's recent trouble spots. The lesson is clear. If a country wants to live at peace with its neighbours its existence must be made vital to those neighbours either as a buffer state or as a producer of some essential commodity. China's long-standing policy of total self-sufficiency and isolationism undoubtedly contributes to her great strength but it also makes her vulnerable. So long as she fails to make other countries depend on her for trade she remains a dispensable threat. At the other extreme the country that is dependent on many others but has little to offer herself is in a position of great weakness.

While it becomes increasingly important to learn to live in harmony with your 'enemies' it is vital not to lose sight of your friends, and this is already a pitfall that we have fallen into in the petroleum era. Mechanization of agriculture has resulted in a trend for fewer and fewer people working on the land whilst there has been rapid expansion of urban populations. Town development has been haphazard, it has followed economic growth patterns rather than social planning. Job structure and the clumping of industrial development, together with price inflation and housing shortages, have resulted in most people residing far from their place of work and also away from their family community.

Ever since man first developed fire and moved out of the tropical belt he has created his own physical and social environment – an environment that is strangely similar in the Arctic or the Tropics, whether he makes his living as hunter-gatherer, agriculturalist, forest dweller, seashore inhabitant, industrial urbanite or whatever. Wherever he is, man always moulds his own micro-environment around himself, the micro-habitat of a *tropical*, *community-living*, *terrestrial ape*.

Men have always occupied fixed or mobile homes which afforded protection from the local climate, which were the focal point of family life, the hearth and home, a place to sleep, somewhere safe to store one's worldly belongings. Groups of homes were built by neighbouring families, kith and kin. Neighbouring villages were more loosely linked by intermarriage, for it was always common practice for a young man to leave his own community to find status and a bride elsewhere before returning to settle in the community of his birth.

In their daily relationships our forebears were surrounded by the familiar faces of neighbours and relatives. Aunts and uncles were at hand to help care for the young. Everyone had grown up within the community so that all the members had known one another since childhood and had shared many joys and sadnesses. Everyone knew whom he could trust or whom to ask for help or advice. This pattern of community life was not limited to small villages. Even in populous oriental and occidental towns the same organization was evident. Chinese businesses were, and are still, essentially family affairs and tight-knit street communities have developed in the Western industrial towns of the last two centuries.

In the last few years, however, everything has changed. New housing estates are occupied by young married couples, the whole estate is a conglomeration of strangers, and friendship and trust grow only slowly out of shyness and suspicion. The children on new estates meet only their own age-peers; they enjoy little intercourse with adolescents or aged members of society. By the time they are teenagers themselves their parents have probably stopped breeding and they have little experience of babies. They have failed to develop strong community relationships and are only too keen to escape from parental authority and set up elsewhere as strangers themselves. In bed-sit land they find themselves among other lonely people, all desperate for love and social acceptance, but all afraid to open themselves up to the strangers around them.

We have so arranged our modern towns that each individual makes far more negative social interactions, cut-off, being ignored or handled brusquely, than they do friendly encounters. It is no wonder that we feel so lost, so uncertain of our own identity, our own acceptability and our own place in society. Nor is it a merely subjective impression that people living in industrial towns today are more miserable, lonely and selfish than their parents and grandparents. Medical statistics show that the number of suicides, personality disorders, nervous breakdowns, psychosomatic symptoms and coronary conditions are all increasing in spite of greater material wealth and improved medical care. Levels of social disorder, vandalism, juvenile crime, muggings, violent crime and terrorism are increasing, while the incidence of divorce, child battering, wife beating, old folk living alone and sexual problems also seem more common. Surely these symptoms are all correlated with the abnormality of the structure of our new communities.

Not only have we failed to preserve the traditional social elements of our environment but we have also failed to preserve the physical elements of the home and the community. The sky-rise tenement with its vertical 'streets' was hailed as the concrete answer to urban housing problems, but families have found such structures uninhabitable. The levels of vandalism by children brought up in many of these tower-blocks have forced councils to close some completely, and in America these symbols of a brave new world are being physically ripped down. High-rise buildings may make ideal office blocks or hotels for a temporary stay, but as terrestrial apes we cannot accept them as a home. Research has shown that even in more conventional housing complexes the actual shape of the buildings and their spatial relationships one to another are enormously important in determining the nature of social relationships that will develop among the strangers who come to live there.

Man may have few instinctive or innate behavioural responses but that does not make him totally plastic. We are born with a very definite design already laid down in our nervous system. Within the limits of normal environments we learn in a fairly predictable pattern. We have the potential to acquire almost limitless cultural details but all human societies share the same few basic cultural themes. Every piece of our verbal language may be culturally learned, but the basic capacity to acquire grammatical language is already laid down in our genes, and our brain is pre-programmed to attend to and deduce the rules from the language we hear around us.

Our bodies, minds, organs and glands are designed to fulfil functions that have proved important to our ancestors, functions that we can call part of human nature, functions that are only appropriate within our traditional environment. If we try to make man adapt in ways foreign to his nature, try to make him fulfil functions for which he is ill-designed, then we put him under great psychological stress. The social planners of the future must keep this in mind; they must define more clearly what are the limits of the natural human environment and then ensure that we are not forced by social, technological or economic progress to live outside these limits.

We are outbreeding our natural resources and our food supplies and this inevitably leads to decreased standards of living and increased social stress and competition. If things were allowed to run their natural course there would be a drastic but effective solution. The powerful and high-ranking individuals would fight to obtain all they needed, whilst there would be high mortality among the lower-ranking have-nots until the population levels had dropped back below the carrying capacity of the environment. This is precisely what does happen among overcrowded populations of rats and mice or even monkeys. Such a solution, however, would be quite contrary to contemporary social attitudes that favour a more even sharing of resources. Nevertheless a combination of such perfect socialism with ever-increasing population size would ultimately lead to everyone getting less and less until the point where we all starved together.

The only way to avoid such predictable disaster is for our species to deliberately check its levels of reproduction. There are several ways in which this could be achieved. There could be legislation with regard to the number of children a couple were allowed to bear. Alternatively all breeding could be made illegal without earned breeding licences and such licences could be awarded only for good services to the community. Breeding could be controlled with each individual being selected for his or her breeding suitability as we do with our domestic cattle. All these methods involve complete infringement of the individual's liberty and most people would regard them with horror, but there are other ways in which reproductive levels can be reduced voluntarily.

By applying appropriate economic and social pressures we could make it more desirable and rewarding to have smaller families. In western societies where population growth is being checked it is due to precisely such social and economic pressures, but there are drawbacks to a policy of small families for all. If we want co-operative, socially responsible citizens we are more likely to rear them in large families than in small ones where selfish attitudes are more easily fostered. It would probably be more satisfactory, and more efficient, to have half as many families breeding but with twice as many children in each. This could be achieved so long as those couples who are keen to have children are not made to feel selfish and those who are not so keen to breed are respected for their usefulness to the community in other ways. When the single-celled organisms first became socialized into multi-cellular organisms, they entrusted the reproductive function to specialized cells, and when individual marine bryozoans became socialized into co-operative colonial creatures they did so by entrusting reproduction to specialist individuals. In the truly social world of the bees, wasps, ants and termites, breeding is again left in the hands of a few society members. It seems quite reasonable that we, too, could achieve far greater social co-operation within our own species by thinking more in terms of population gene pools rather than individuals and entrusting the responsibility of rearing future members of our communities to breeding specialists.

In man, as in other animals, natural selection has favoured the pursuit of goals that have aided the individual in his evolutionary function of breeding and rearing young successfully. Such goals have included the pursuit of wealth and status, courtship, sex and friendly social relationships. Yet the gratification of man lies largely in his achieving these goals irrespective of whether they ultimately contribute to breeding or not. I would like to imagine that my own lineage will continue after my lifespan but whether it dies out or not is really no matter of great concern in my life. I am far more anxious to know where my next pay cheque is coming from, whether I am going to get my potato crop up before the first frost, what we are going to eat for dinner tonight and whether we should have a large or a small Christmas tree this year. I have two sons and their welfare is of great interest to me; but whether they decide to become celibate monks or polygamous superstuds is their concern, not mine.

There are a multitude of worthwhile and fulfilling goals for our non-breeding community members, whether these are self-indulgent like the expensive cruise and bigger car, altruistic community service or a mixture of both. In the more stimulating industrial societies men need little encouragement to pursue selfish goals rather than put all their efforts into breeding, and in the most advanced industrial countries population growth has already dropped to a near zero figure. If similarly attractive alternatives to breeding were made available to the less developed areas of the globe and there could be a shift from selfish to more altruistic goals, mankind would be well on its way towards increased social co-operation and reduced competition, whilst still maintaining the qualities we appreciate in life.

Altruism is hard to sell. It is difficult to persuade workers to restrain their pay demands for the good of the community, as we are finding out in Britain today. A few brainwashed mugs might do as they are told but most people will only support such a demand if they can see that they will benefit from the community in some other way. The pay-offs for such altruism can be in respect, status, increased power or a place in the history books rather than in hard cash or reproductive success. Wealthy North American Indians showed enormous generosity in their potlatch gift-giving ceremonies because they were accorded respect and status proportionate to their expense. Similarly celibate religious orders have always been able to persuade recruits that their enhanced social standing will more than outweigh the reproductive futility of their vows.

Human contentment or satisfaction derives from the successful achievement of realistic goals and the receipt of appropriate respect and acceptance from our fellows. Yet somehow this simple concept of happiness seems increasingly unattainable in our modern world of fast-changing social attitudes.

As mankind moves steadily towards greater social equality and international racial harmony, racial, sexual and class barriers are weakened and broken down. Jack becomes as good as his master and outstanding wealth or enterprise are no longer treated with respect but despised as anti-social exploitation, or showing up one's fellows. In several Western countries the disruption of community structure and discipline, decreasing credibility of religious dogma, changing social power structures and liberal education systems all contribute to this collapse of respect and responsibility.

How can Mr Average gain respect when he can no longer feel pride in his race, his class, his position in society, his particular talents, his workmanship or his modest achievements? He can never attain the elegant style of life that the advertising media promise, as these goals are unrealistic for the majority. He feels bitter and frustrated. He does not really need the cars and speedboats, cocktail bars and swimming pools nor all the other gadgetry of modern stylish living to be happy, but he does not like to see other folk getting a better deal than himself. If material possessions are the only symbols of status and respect he can achieve, then he will selfishly pursue them. There is clearly a great deal of re-adjustment to be made to our social values of goals and rewards before Mr Average can feel content with his lot in this crowded society.

And Ms Average, is she content with her foolproof contraceptives, her sexual emancipation and her full education? She is still less able than her male colleagues at performing many physical tasks but in industrialized communities at least there are plenty of fields where she can compete on equal terms. With fewer women breeding and those that do having fewer children and spending less time in motherhood it is surely right that women should be granted the equal opportunity of employment outside the home. Nevertheless Ms Average is a woman, not a man. She is not so aggressive, she is probably more emotional and her biological make-up has prepared her, indeed designed her, for fulfilling quieter, less spectacular roles than those demanded by the more competitive status-seeking males. For all her opportunity and capability Ms Average is probably going to end up in a supportive domestic role. But now that she is liberated she has not only lost many of her previous privileges but she may feel unfulfilled and haunted by the misery that she ought to be making more of herself.

Women's libbers will never achieve social parity with men through equal opportunity. They will be constantly let down by their sex's biologically lower motivation for fighting for glory in the industrial head-hunt of the economic rat-race. Parity will only be achieved when the domestic and maternal roles of women are given their proper social value. The very fact that our species is typified by a one to one sex ratio in the adult breeding population shows that in evolutionary terms the male and female roles are of equal value. When the female roles are awarded the respect and remuneration they deserve friction between the sexes will be largely avoided.

Only if we avoid destroying ourselves and our environment is it meaningful to consider the future for the other members of our ape superfamily. But assuming we can pull through, what prospects can these rare and endangered creatures have? The survival chances of the orang-utan, mountain gorilla and some species of gibbon are already slim and all the other species will soon be in danger if we continue to cut down forests at the present rate.

Can the modern apes adapt to new forest-free environments as our own ancestors once did? The answer is almost certainly no. Our ancestors evolved during a long period of very gradual change from rainforest through seasonal monsoon forest to savannah woodlands. The whole process took several million years and even then we were probably lucky to have come through at all. What hope do the present forest apes have of adapting to a similar environmental change in a couple of generations? Only the chimpanzee shows any signs of being able to live in open country but with the whole Sahelian region south of the Sahara becoming more arid every year suitable chimp habitat will soon be rare. Our efforts at conservation may protect breeding populations of some apes within reserved areas of their natural habitat but the very activity of preservation or game management aims at maintaining a desirable *status quo* and there will be little opportunity for further evolutionary advance by apes surviving in such reserves.

Captive populations of apes are in a very different position, however. Apes are not difficult to breed in zoos and laboratories and are already used extensively in medical research and psychological research. It is sixteen years since the first chimpanzee explored space. Apes are intelligent animals that can be trained to perform complex tasks with great reliability. We can even talk with some of our captive chimpanzees. Man has found uses for many large intelligent animals. Hannibal employed elephants as ancient war machines and these grand animals are still used for transport and traction work. Marine mammals such as dolphins have proved indispensable in submarine spying operations and can even be trained to place limpet mines on enemy vessels or retrieve valuable objects from the sea bed. Man is sure to apply selective breeding on all these useful creatures to exaggerate the features he finds most valuable to him, just as he has with the dog and horse. If the vagaries of natural selection could triple man's brain size in three million years one wonders what the more potent pressures of artificial selection could do to the chimp brain in a thousand generations. The possibility of breeding races of sub-human apes to perform a variety of useful tasks is only too real and too probable.

Imagine breeds of enormously strong gorillas, intelligent chimps and patient orang-utans doing our work and running many of our essential services. Imagine them one day demanding equal rights, maybe taking over their own management, or even the whole world, as in Pierre Boulle's *Monkey Planet.* It all seems very far-fetched but within the living apes lies a remarkable nucleus of genetic material, very similar to that from which the right selection pressures wrought our own humanoid condition.

If this all sounds like science fiction it is not surprising. Science fact of today is the science fiction of yesterday and the pace of scientific progress is increasing all the time. Men may soon be building submarine and subterranean cities, fertilizing the deserts and melting the polar ice caps. Scientists may develop photosynthesis units to convert sunlight directly into edible sugars without our having to rely on delicate, ephemeral, disease-prone plants to do the job for us. Mankind will find better ways to tap the enormous natural energy of our planet, its spin, its heat, the pull of the moon, the movements of the winds and seas. Man will be playing with genetic material like a child with building blocks, designing new animals, new plants, new drugs, new diseases.

Space beckons our intelligent species like the singing sirens who lured so many ancient sailors to their deaths on the Aegean rocks. We are unable to resist its mystery and already our interplanetary ships are probing the possibilities for human life in other parts of the universe. It might well be more sensible to ignore space and concentrate on improving the quality of life on our own sweet planet but that would be quite out of character. We are the species that burst blindly out of our African cradle into the unknown forests of Asia as soon as there was a land bridge, the species that turned its back on the security of tropical life to face the harsh northern winters, the bears and the wolves. We are the species that flaunted the impossible living conditions of three major Ice Ages, the species that walked across the frozen Bering Strait on the off-chance that there was land on the other side. It was our own supposedly intelligent species that set sail into the unknown blue to colonize Polynesia, Micronesia and Australasia. Even when they believed the earth was flat and that they would all fall off the edge, Columbus's crew did not fear to sail across uncharted seas, just as the Vikings had done several centuries earlier. Is it really likely that we will resist those real and certain stars that wink so invitingly every night? Surely somewhere among those billions of suns and planets lie the Utopian dreams that drive our species so furiously onwards.

When, like the ancient Maya, we have fouled and raped our own world to the limit we will have to turn to space to find new empires. The upright ape, who evolved a strange brainy intelligence but never stayed still long enough to acquire much wisdom, will embark on his galactic adventures with as little idea of where his novel actions will lead as had the first fish that crawled out of the Devonian sea some four hundred million years ago.

APPENDIX

Man and Apes – Some Vital Statistics

		Man	Chimp	Gorilla	Orang-utan	Gibbon	Siamang
Number of chromosomes		46	48	48	48	44	50
Average newborn weight in kilograms		3·29	1·58	1·92	1·48	0·40	0·56
Gestation period in days		275	242	265	254	210	230
Average age at sexual maturity in years		14	7–8	7–8	7–8	6–8	7–8
Average adult weight in kilograms	male	65	45	175	75	5·7	11·1
	female	58	40	85	57	5·3	10·3
Average cranial capacity in cubic centimetres	male	1400	396	535	414	104	160
	female	1300	355	458	366	101	120

Bibliography

INTRODUCTION

1. Schaller, G. B., 1963, The Mountain Gorilla. University of Chicago Press, Chicago.
   Schaller, G. B., 1964, The Year of the Gorilla. Collins, London.
2. Fossey D., 1972, Vocalizations of the Mountain Gorilla (*Gorilla gorilla beringei*). *Anim. Behav.*, 20.
   Fossey, D., 1974, Observations on the Home Range of One Group of Mountain Gorillas (*Gorilla gorilla beringei*). *Anim. Behav.*, 22.
3. Lawick-Goodall, J. van, 1968, The Behaviour of Free-Living Chimpanzees in the Gombe Stream Reserve. *Anim. Behav. Monogr.*, 1.
   Lawick-Goodall, J. van, 1971, In the Shadow of Man. Collins, London.
4. Carpenter, C. R., 1940, A Field Study in Siam of the Behaviour and Social Relations of the Gibbon (*Hylobates lar*). *Comp. Psychol. Monogr.*, 16 (5).
5. Ellefson, J. O., 1974, A Natural History of White-handed Gibbons in the Malayan Peninsula. In 'Gibbon and Siamang', vol. 3. Karger, Basel.
6. Chivers, D. J., 1974, The Siamang in Malaya; A Field Study of a Primate in Tropical Rainforest. *Contrib. Primat.*, vol. 4. Karger, Basel.
7. Horr, D. A., 1975, The Borneo Orang-utan; Population Structure and Dynamics in Relationship to Ecology and Reproductive Strategy. In 'Primate Behaviour, Developments in Field and Laboratory Research', vol. 4. L. A. Rosenblum (ed.). Academic Press, London.

8. Rodman, P. S., 1973, Population Composition and Adaptive Organization among Orang-utans of the Kutai Reserve. In 'Comparative Behaviour and Ecology of Primates'. R. P. Michael and J. H. Crook (eds.). Academic Press, London.

9. Rijksen, H. D., 1975, Social Structure in a Wild Orang-utan Population, Sumatra. *Contemp. Primat. 5th Int. Congr. Primat.* Karger, Basel.

10. Galdikas-Brindamour, B., 1978, Orang-utan Adaptation at Tanjong Puting Reserve: Mating and Ecology. In 'Perspectives on Human Evolution', vol. 4. D. Hamburg (ed.). Addison-Wesley, USA.

11. Sabater Pi, J., 1974, An Elementary Industry of the Chimpanzees in the Okorobiko Mountains, Rio Muni (Republic of Equatorial Guinea), West Africa. *Primates*, 15 (4).

12. Harcourt, A. H., Stewart, K. J. and Fossey, D., 1976, Male Emigration and Female Transfer in Wild Mountain Gorilla. *Nature*, 263.

13. Kortlandt, A., 1965, How Does a Chimpanzee Use Weapons When Fighting Leopards? *Yearb. Amer. Phil. Soc.*, 1965.

Kortlandt, A., 1972, New Perspectives on Ape and Human Evolution. *Stichting voor Psychobiologie*. University of Amsterdam.

14. Reynolds, V., 1965, Budongo: A Forest and its Chimpanzees. Natural History Press, New York.

15. Badrian, A. and N., 1977, Pygmy Chimpanzees. *Oryx*, 13 (5).

16. Sugiyama, Y., 1973, The Social Structure of Wild Chimpanzees: A Review of Field Studies. In 'Comparative Behaviour and Ecology of Primates'. R. P. Michael and J. H. Crook (eds.), Academic Press, London.

Bibliography

CHAPTER 1

1. MacKinnon, J. R. and K. S., 1977, Comparative Feeding Ecology of Six Sympatric Malaysian Forest Primates. In '*Proc. 6th Int. Congr. Primat.*'. Academic Press, London.
2. Chivers, D. J., 1974, The Siamang in Malaya: A Field Study of a Primate in Tropical Rainforest. *Contrib. Primat.*, vol. 4. Karger, Basel.
   Chivers, D. J., Raemaekers, J. J. and Aldrich-Blake, F. P. G., 1975, Long-Term Observations of Siamang Behaviour. *Folia primat.*, 23.
3. MacKinnon, J. and K., 1977, The Formation of a New Gibbon Group. *Primates*, 18 (3).

CHAPTER 2

1. Fleagle, J. G., 1976, Locomotion and Posture of the Malayan Siamang and Implications for Hominoid Evolution. *Folia primat.*, 26.
2. Chivers, D. J., Raemaekers, J. J. and Aldrich-Blake, F. P. G., 1975, Long-Term Observations of Siamang Behaviour. *Folia primat.*, 23.
3. Chivers, D. J. and MacKinnon, J., 1977, On the Behaviour of Siamangs after Playback of Their Calls. *Primates*, 18 (4).
4. MacKinnon, J., 1977, A Comparative Ecology of Asian Apes. *Primates*, 18 (4).
5. Chivers, D. J., 1971, Spatial Relations Within the Siamang Group. In '*Proc. 3rd Int. Congr. Primat.*', vol. 3. H. Kummer (ed.). Karger, Basel.
6. Tenaza, R. R., 1975, Territory and Monogamy Among Kloss' Gibbons (*Hylobates klossi*) in Siberut Island, Indonesia. *Folia primat.*, 24 (1).
7. Deag, J. M. and Crook, J. H., 1971, Social Behaviour and 'Agonistic Buffering' in the Wild Barbary Macaque (*Macaca sylvana*). *Folia primat.*, 15.

CHAPTER 3

1. Casimir, M. J. and Butenandt, E., 1973, Migration and Core Area Shifting in Relation to Some Ecological Factors in a Mountain Gorilla Group (*Gorilla gorilla berengei*) in the Mount Kahuzi Region (République du Zaïre). *Zeit. für Tierpsychol.*, 33.
2. Schaller, G. B., 1963, The Mountain Gorilla. University of Chicago Press, Chicago.
   Schaller, G. B., 1964, The Year of the Gorilla. Collins, London.
3. Fossey, D., 1972, Vocalizations of the Mountain Gorilla (*Gorilla gorilla berengei*). *Anim. Behav.*, 20.
   Fossey, D., 1974, Observations on the Home Range of One Group of Mountain Gorillas (*Gorilla gorilla berengei*). *Anim. Behav.*, 22.
4. Jones, C. and Sabater Pi, J., 1971, Comparative Ecology of *Gorilla gorilla* (Savage and Wyaman) and *Pan troglodytes* (Blumenbach) in Rio Muni, West Africa. *Bibl. Primatol.*, 13.

CHAPTER 4

1. McGinnis, P., 1973, Sexual Behaviour of Chimpanzees. Ph.D thesis, Cambridge University.
2. Bygott, D., 1972, Cannibalism Among Wild Chimpanzees. *Nature*, 238.
3. Goodall, J., 1963, Feeding Behaviour of Wild Chimpanzees. *Symp. Zool. Soc. Lond.*, 10.
4. Lawick-Goodall, J. van, 1968, The Behaviour of Free-Living Chimpanzees in the Gombe Stream Reserve. *Anim. Behav. Monogr.*, 1.
5. Jones, C. and Sabater Pi, J., 1969, Sticks Used by Chimpanzees in Rio Muni, West Africa. *Nature*, 223.
6. Sabater Pi, J., 1974, An Elementary Industry of the Chimpanzees in the Okorobiko Mountains, Rio Muni

(Republic of Equatorial Guinea), West Africa. *Primates* 15 (4).

7. Goodall, J., 1964, Tool-Using and Aimed Throwing in a Community of Free-Living Chimpanzees. *Nature*, 201.
8. Kortlandt, A., 1965, How Does a Chimpanzee Use Weapons When Fighting Leopards? *Yearb. Amer. Phil. Soc.*, 1965.
9. Goodall, J., 1962, Nest Building Behaviour in Free-Ranging Chimpanzees. *Ann. N. Y. Acad. Sci.*, 102.

CHAPTER 5

1. Badrian, A. and N., 1977, Pygmy Chimpanzees. *Oryx*, 13 (5).
2. MacKinnon, J., 1976, Mountain Gorillas and Bonobos. *Oryx*, 13 (4).
3. Tulpius, N., 1641, *Observationummedicarum libri tres*, pp. 274–9. Amsterdam.
4. Reynolds, V., 1968, The Apes. Cassell, London.

CHAPTER 6

1. MacKinnon, J., 1974, The Behaviour and Ecology of Wild Orang-utans (*Pongo pygmaeus*). *Anim. Behav.*, 22. MacKinnon, J., 1974, In Search of the Red Ape. Collins, London.
2. Galdikas-Brindamour, B., 1978, Orang-utan Adaptation at Tanjong Puting Reserve: Mating and Ecology. In 'Perspectives on Human Evolution', vol. 4. D. Hamburg (ed.). Addison-Wesley, USA.
3. MacKinnon, J., 1977, Reproductive Behaviour in Wild Orang-utan Populations. In 'Perspectives on Human Evolution', vol. 4. D. Hamburg (ed.). Addison-Wesley, USA.
4. Tuttle, R. H., 1969, Knuckle-Walking and the Problem of Human Origins. *Science*, 166.

CHAPTER 7

1. Darwin, C., 1859, On the Origin of Species by Means of Natural Selection. Murray, London.
   Darwin, C., 1871, The Descent of Man and Selection in Relation to Sex. Murray, London.
2. Keith, A., 1923, Man's Posture: Its Evolution and Disorders. *Brit. Med. J.*, 1.
3. Pilbeam, D., 1970, The Evolution of Man. Thames and Hudson, London.
4. Goodman, M. and Moore, G. W., 1971, Immunodiffusion Systematics of the Primates, I. The Catarrhini. *Systemat. Zool.*, 20.
5. Chiarelli, B., 1966, Caryology and Taxonomy of the Catarrhine Monkeys. *Am. J. Phys. Anthrop.*, 24.
6. Sarich, U. M. and Wilson, A. C., 1967, Immunological Time Scale for Hominid Evolution. *Science*, 158.
7. Kortlandt, A., 1972, New Perspectives on Ape and Human Evolution. *Stichting voor Psychobiologie*. University of Amsterdam.
8. Romero-Herrera, A. E., Lehmann, H., Costillo, O., Joysey, K. A. and Friday, A. E., 1976, Myoglobin of the Orang-utan: A Phylogenetic Enigma. *Nature*, 261.
9. Howell, F. C., 1965, Early Man. Time-Life Books.
10. Jolly, C. J., 1970, The Seed-Eaters: A New Model of Hominid Differentiation Based on a Baboon Analogy. *Man, N. S.*, 5.
11. Martin, P. S., 1966, Africa and Pleistocene Overkill. *Nature*, 212.
12. Morris, D., 1967, The Naked Ape. Cape, London.
13. Morgan, E., 1972, The Descent of Woman. Souvenir Press.
14. Napier, J. R., 1964, The Evolution of Bipedal Walking in the Hominids. *Arch. Biol.* (*Liège*) *75: Suppl.*

CHAPTER 8

1. Lee, R. B. and DeVore, I. (eds.), 1968, Man the Hunter. Aldine, Chicago.
2. Woodburn, J., 1968, An Introduction to Hadza Ecology. In 'Man the Hunter'. R. B. Lee and I. DeVore (eds.). Aldine, Chicago.
3. Yerkes, R. M., 1925, Traits of Young Chimpanzees. In 'Chimpanzee Intelligence and its Vocal Expression'. R. M. Yerkes and B. W. Learned (eds.). Williams and Wilkins, Baltimore.
4. Hayes, C., 1951, The Ape in our House. Harper and Row, New York.
5. Premack, A. J. and D., 1972, Teaching Language to an Ape. *Scientific American*, 227.
6. Rumbaugh, D., Gill, T. V. and von Glaserfeld, E. C., 1973, Reading and Sentence Completion by a Chimpanzee (*Pan*). *Science*, 182.
7. Gardner, R. A. and B. T., 1969, Teaching Sign Language to a Chimpanzee. *Science*, 165.
8. Fouts, R., 1973, Acquisition and Testing of Gestural Signs in Four Young Chimpanzees. *Science*, 180.
9. Menzel, E. W., 1971, Communication About Environment in a Group of Young Chimpanzees. *Folia primat.*, 15.
10. Lips, J. E., 1949, The Origin of Things: A Cultural History of Man. Harrap, London.
11. Freeman, J. D., 1960, Bird Augury. In 'Birds of Borneo'. B. E. Smythies, Oliver and Boyd, Edinburgh and London.
12. Morris, D., 1962, The Biology of Art. Methuen, London.

CHAPTER 9

1. Blurton-Jones, N. G., (ed.), 1977, Ethological Studies of Child Behaviour. Cambridge, New York and London.

2. Rijksen, H. D., 1975, Social Structure in a Wild Orang-utan Population, Sumatra. *Contemp. Primat. 5th Int. Congr. Primat.* Karger, Basel.

CHAPTER 10

1. Trivers, R. L., 1971, The Evolution of Reciprocal Altruism. *Q. Rev. Biol.*, 46.
2. Rijksen, H. D., 1977, Origins of Hunting in Hominids. In *Proc. 6th Int. Congr. Primat.* Academic Press, London.
3. Reynolds, V., 1966, Open Groups in Hominid Evolution. *Man*, 1 (4).
4. Ardrey, R., 1961, African Genesis. Collins, London.
   Ardrey, R., 1966, The Territorial Imperative. Collins, London.
   Ardrey, R., 1976, The Hunting Hypothesis. Collins, London.
5. Lorenz, K., 1966, On Aggression. Methuen, London.
6. Montagu, A., 1976, The Nature of Human Aggression. Oxford University Press, New York.
7. Kenyatta, J., 1938, Facing Mount Kenya. Secker and Warburg, London.

Index